# CONTROL SYSTEMS

## Analysis and Realization

**Jesus C. de Sosa**

CONTROL SYSTEMS
Analysis and Realization

*iUniverse books may be ordered through booksellers or by contacting:*

*iUniverse*
*1663 Liberty Drive*
*Bloomington, IN 47403*
*www.iuniverse.com*
*1-800-Authors (1-800-288-4677)*

*Because of the dynamic nature of the Internet, any Web addresses or links contained in this book may have changed since publication and may no longer be valid. The views expressed in this work are solely those of the author and do not necessarily reflect the views of the publisher, and the publisher hereby disclaims any responsibility for them.*

*ISBN: 978-1-4502-1607-4 (sc)*
*ISBN: 978-1-4502-1608-1 (ebook)*

*Printed in the United States of America*

*iUniverse rev. date: 4/8/2010*

For my grandchildren,

Anthony Diesta,

Joseph Diesta,

Mia DeJesus, &

Miley DeJesus,

who gave me thousands of smiles.

# TABLE OF CONTENTS

# List of Tables

# List of Figures

# Preface

Control system is exciting. It encompasses all electronic systems. A typical electronic system consists of a power supply, communication, and RF/microwave subsystems. Each subsystem has one or more elements of control system. For example, a power supply uses a fast switch to convert a 60-Hz power system to a DC power supply. Similarly, communication subsystem employs a phase lock loop, which is essentially a control block. A tracking receiver for a RF/microwave subsystem is also a control block.

Due to its wide applications, the basics of control system must be understood. This book is not a textbook on control systems. It is an illustration book that emphasizes the basics. While a textbook emphasizes rigor in representing a concept, the book illustrates the concept. As a result, the book uses extensive figures and tables to illustrate the basic concepts of control systems.

Chapter 1 discusses the important role of high school algebra in understanding the basics. Expressions and equations of control systems must be represented into their standard forms.

Transfer function provides the unifying role in the analysis for stability and performance of a control system. In Chapter 2, the author showed how to use node admittance matrix to derive the transfer function of a system.

Chapter 3 introduces block diagrams. A block diagram is a graphical language of control systems. Several rules in simplifying a complex block diagram are shown. The chapter also includes a method of deriving the transfer function of a block diagram.

Signal flow graph is another graphical language of control system. Chapter 4 introduces signal flow graphs, their construction, and their analysis using Mason's rule.

The matrix approach in solving a system of differential equations is one of the most general and important method of solving problems in control systems. It is, however, a more advanced approach and best applied with computer programming. Chapter 5 introduces the essence of the approach. Several examples show the details of the approach.

Chapter 6 shows Bode plots. A Bode plot shows the gain and phase of a transfer function's constants, simple poles and zeros, and quadratic poles and zeros. The chapter also shows how to construct the composite plot of all the gains and the phases.

A quadratic factor causes overshoot time, rise time, and settling time. Chapter 7 analyzes a quadratic pole to explain the above parameters. These parameters are crucial not only in the design of a control system but in its tests as well.

Chapter 8 shows how a realization diagram may be constructed. A realization diagram is not a design diagram. The former describes in a generic way the parts of a control system. Design diagram are sketches or drawings that show how to build a control system. A realization diagram is the bridge between the mathematics of a control system and its design diagrams.

A realization diagram is not complete without analyzing for its stability. Chapter 9 employ the Routh-Hurwitz stability criterion and Bode plot in analyzing the stability of a system. Additionally, Chapter 9 also introduces time scaling and amplitude scaling. They are useful when modifying a realization diagram.

The last chapter, or chapter 10, introduces difference or recursion equations and their corresponding z-transforms. Z-transform is to digital while Laplace transform is to analog. The chapter also shows examples of realization diagrams of digital control systems.

# Chapter 1

# Algebra in Control Systems

For the most part, the study of control system requires high school algebra as the main mathematical tool. This is made possible because of Laplace transform. Laplace transform converts the differential equation of electrical circuits into algebraic equations. The latter equations are easier to solve than differential equations.

While Laplace transform is used in continuous systems, z-transform is used in discrete systems. The two transforms are related in the sense that z-transform can be derived from Laplace transform.

After defining Laplace transform and z-transform, this chapter reviews the basics of high school algebra from manipulating expressions to solving simultaneous equations.

## 1.1    THE LAPLACE TRANSFORM

The Laplace transform is defined by

$$F(s) = \int_0^\infty f(t)e^{-st}\,dt \qquad t \geq 0$$

where

$s$ = complex frequency variable, and

$t$ = time variable.

Inspecting the above definition of Laplace transform shows that integration is required in finding the Laplace transform of $f(t)$. The variable of integration is the time, $t$. It has one-sided limits from zero to infinity. Note that the Laplace variable $s$ is a parameter during the integration of $F$.

Example 1. Find the Laplace transform of $f(t) = 1$.

Using the definition,

$$F(s) = \int_0^\infty (1)\left(e^{-st} dt\right).$$

Carrying the integration,

$$F(s) = -\frac{1}{s} e^{-st} \Big|_0^\infty$$

or,

$$F(s) = \frac{1}{s} \qquad s \neq 0.$$

Example 2. Find the Laplace transform of $f(t) = e^t$.
Doing the same procedure as in example 1,

$$F(s) = \int_0^\infty \left(e^t\right)\left(e^{-st} dt\right).$$

Simplifying the exponential term,

$$F(s) = \int_0^\infty e^{(1-s)t} dt$$

Using integration by parts, let

$$u = (1-s)t$$

$$du = 1-s$$

$$\int e^{(1-s)t} dt = \int e^u du$$

After substituting the limits of integration, the Laplace transform becomes

$$F(s) = \frac{1}{s-1} \qquad s \neq 1.$$

Note the limitations on $s$. Values of $s$ that make the denominator of $F(s)$ zero are not allowed. That is, when the denominator is zero, $F(s)$ is infinite. For the first example, $s$ cannot be zero. The second example allows $s$ to be zero but cannot be one.

Table 1.1 shows the Laplace transform of common functions. In the design of electronic systems, certain input and output signals are required. These signals are function of time. To find their Laplace transform, simply look at a table similar to Table 1.1 and find their transforms.

Table 1.2 generalizes some of the most important properties of Laplace transform. The linearity property shows that the constants $a$ and $b$ remain the same before and after getting the Laplace transform of a sum.

Other usefulness of Table 1.2 involve the quick determination of a Laplace transform when the time variable, $t$, is shifted or scaled. Example of shifting of $t$ is $t - \tau$, which essentially means the original $t$ becomes $t - \tau$. An example of scaling is $at$. Here, the original $t$ is replaced by $at$. By using Table 1.1 with Table 1.2, other Laplace transforms pairs may be derived.

Table 1.1 Laplace Transform of Commonly used Functions in Engineering

| Description of the function | Function of time $f(t)$ | Laplace transform |
|---|---|---|
| Ideal delay | $\delta(t-\tau)$ | $e^{-\tau s}$ |
| Unit impulse | $\delta(t)$ | $1$ |
| Unit step | $u(t)$ | $\dfrac{1}{s}$ |
| Delayed unit step | $u(t-\tau)$ | $\dfrac{e^{-\tau s}}{s}$ |
| Ramp | $t \cdot u(t)$ | $\dfrac{1}{s^2}$ |
| Exponential decay | $e^{-at}u(t)$ | $\dfrac{1}{s+a}$ |
| Exponential approach | $\left(1-e^{-at}\right)u(t)$ | $\dfrac{a}{s(s+a)}$ |
| Sine | $\sin(\omega t)u(t)$ | $\dfrac{\omega^2}{s^2+\omega^2}$ |
| Cosine | $\cos(\omega t)u(t)$ | $\dfrac{s}{s^2+\omega^2}$ |

Table 1.2 Some Properties of the Laplace Transform

| Property | Function of time $f(t)$ | Laplace transform |
|---|---|---|
| Linearity | $af(t) + bg(t)$ | $aF(s) + bG(s)$ |
| Frequency differentiation | $tf(t)$ | $-F'(s)$ |
| First derivative | $f'(t)$ | $sF(s) - f(0^-)$ |
| Second derivative | $f''(t)$ | $s^2 F(s) - sf(0^-) - f'(0^-)$ |
| Frequency integration | $\dfrac{f(t)}{t}$ | $\displaystyle\int_0^\infty F(\sigma)d\sigma$ |
| Integration | $\displaystyle\int_0^t f(\tau)d\tau = u(t) * f(t)$ | $\dfrac{1}{s}F(s)$ |
| Scaling in time | $f(at)$ | $\dfrac{1}{|a|}F\left(\dfrac{s}{a}\right)$ |
| Shifting in frequency | $e^{at} f(t)$ | $F(s - a)$ |
| Shifting in time | $f(t-a)u(t-a)$ | $e^{-as} F(s)$ |
| Convolution | $(f * g)(t)$ | $F(s)G(s)$ |

## 1.1.1    Laplace transform in electrical engineering

The Laplace transform converts a set of differential equations into a set of algebraic equations involving the Laplace variable $s$. In manual calculations, algebraic equations are easier to solve than a set of differential equations.

Essentially, the Laplace variable $s$ is the ratio of the first derivative of a variable and the variable. Figure 1.1(a) shows how $s$ in a capacitor can be associated with the ratio of the first derivative of voltage across the capacitor and the voltage itself. The derivation starts with the basic definition of current and equating the result to impedance. Similarly, $s$ in an inductor is the ratio of the first derivative of current through an inductor and the current itself. This is shown on Figure 1.1(b).

## 1.2      THE Z-TRANSFORM

The z-transform of a discrete time signal x[n] is defined as

$$X(z) = Z\{x[n]\}$$

$$= \sum_{n=0}^{\infty} x[n]z^{-n}$$

where

$z = Ae^{j\varphi}.$
$A$ = magnitude of z, and
$\varphi$ = angle or phase in radians.

Note that the index of the summation starts with zero.

As an example, consider that A = 1 and $x[n] = 1, 2, 3, ....$ Then, its z-transform is

$$X(z) = 1z^{0} + 2z^{-1} + 3z^{-2}.$$

$$i_C = C\frac{dV_C}{dt}$$

$$\frac{i_C}{V_C} = sC$$

$$i_C = sCV_C$$

$$s = \frac{1}{V_C}\frac{dV_C}{dt}$$

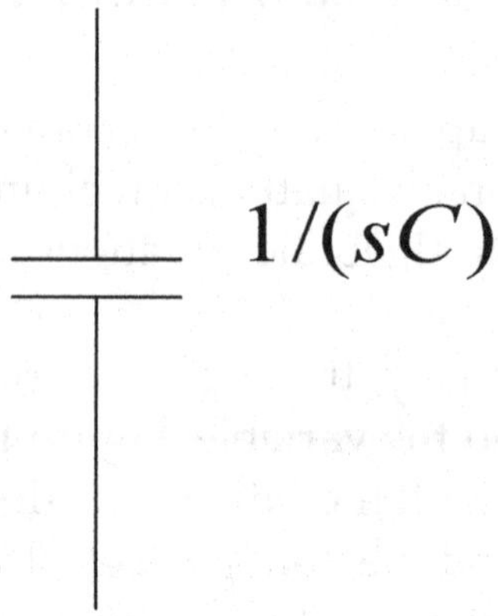

$1/(sC)$

(a) Interpretation of $s$ in a capacitor

$$V_L = L\frac{di_L}{dt}$$

$$\frac{V_L}{i_L} = sL$$

$$V_L = sLi_L$$

$$s = \frac{1}{i_L}\frac{di_L}{dt}$$

$sL$

(b) Interpretation of $s$ in an inductor

Figure 1.1 Interpretation of s in a Capacitor and Inductor

The z-transform uses the same coefficients as x[n]. However, the exponent of $z$ becomes more negative. To illustrate this point, Figure 1.2 shows the variation of the angle as a signal moves around a unit circle on a complex plane.

Table 1.3 shows some important z-transform pairs and Table 1.4 are some of its properties. Getting back x[n] from its z-transform is by the inverse transform,

$$x[n] = Z^{-1}\{X(s)\}$$
$$= \frac{1}{2\pi j} \oint_C X(z) z^{n-1} dz$$

The region of convergence identifies the value of z for which a z-transform is stable. It is defined by

$$ROC = \{z : \left| \sum_{n=-\infty}^{\infty} x[n] z^{-n} \right| < \infty\}$$

Sometimes it is desirable to transform a Laplace transform to z-transform. The following change in variable may be used:

$$s = \frac{2z - 1}{Tz + 1}.$$

Conversely, a z-transform may be converted to Laplace transform via:

$$z = \frac{2 + sT}{2 - sT}.$$

Two other theorems are useful. They are the initial value theorem,

$$x[0] = \lim_{z \to \infty} X(z) \text{ if x[n] is causal, and}$$

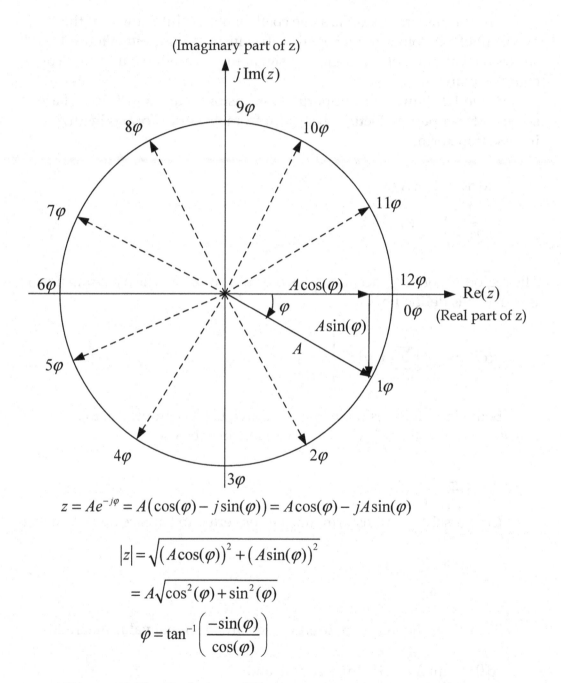

$$z = Ae^{-j\varphi} = A\big(\cos(\varphi) - j\sin(\varphi)\big) = A\cos(\varphi) - jA\sin(\varphi)$$

$$|z| = \sqrt{\big(A\cos(\varphi)\big)^2 + \big(A\sin(\varphi)\big)^2}$$

$$= A\sqrt{\cos^2(\varphi) + \sin^2(\varphi)}$$

$$\varphi = \tan^{-1}\left(\frac{-\sin(\varphi)}{\cos(\varphi)}\right)$$

Figure 1.2 The Variation of Phase Angle in the Complex Plane

Table 1.3 Some z-Transform Pairs

| x[n] | X(z) | Region of convergence |
|---|---|---|
| $\delta[n]$ | 1 | all z |
| $\delta[n-n_0]$ | $z^{-n_0}$ | $z \neq 0$ |
| $u[n]$ | $\dfrac{1}{1-z^{-1}}$ | $|z|>1$ |
| $-u[-n-1]$ | $\dfrac{1}{1-z^{-1}}$ | $|z|<1$ |
| $nu[n]$ | $\dfrac{z^{-1}}{\left(1-z^{-1}\right)^2}$ | $|z|>1$ |
| $a^n u[n]$ | $\dfrac{1}{1-az^{-1}}$ | $|z|>a$ |
| $\cos(\omega_0 n)u[n]$ | $\dfrac{1-z^{-1}\cos(\omega_0)}{1-2z^{-1}\cos(\omega_0)+z^{-2}}$ | $|z|>1$ |
| $\sin(\omega_0 n)u[n]$ | $\dfrac{z^{-1}\sin(\omega_0)}{1-2z^{-1}\cos(\omega_0)+z^{-2}}$ | $|z|>1$ |

## Table 1.4 Some Properties of the z-Transform

| Property | $x[n]$ | $X(z)$ |
|---|---|---|
| Linearity | $a_1 x_1[n] + a_2 x_2[n]$ | $a_1 X_1(z) + a_2 X_2(z)$ |
| Time shifting | $x[n-k]$ | $x^{-k} X(z)$ |
| Scaling in the z-domain | $a^n x[n]$ | $X(a^{-1}z)$ |
| Time reversal | $x[-n]$ | $X(z^{-1})$ |
| Conjugation | $x^*[n]$ | $X^*(z^*)$ |
| Real part | $\mathrm{Re}\{x[n]\}$ | $\dfrac{1}{2}\left[X(z) + X^*(z^*)\right]$ |
| Imaginary part | $\mathrm{Im}\{x[n]\}$ | $\dfrac{1}{2j}\left[X(z) - X^*(z^*)\right]$ |
| Differentiation | $nx[n]$ | $-z\dfrac{dX(z)}{dz}$ |
| Convolution | $x_1[n] * x_2[n]$ | $X_1(z) X_2(z)$ |
| Multiplication | $x_1[n]x_2[n]$ | $\dfrac{1}{j2\pi}\oint_c X_1(v) X_2\left(\dfrac{z}{v}\right) v^{-1} dv$ |
| Parseval relation | $\displaystyle\sum_{n=-\infty}^{\infty} x_1[n]x_2^*[n]$ | $\dfrac{1}{j2\pi}\oint_c X_1(v) X_2^*\left(\dfrac{z}{v^*}\right) v^{-1} dv$ |

the final value theorem:

$$x[\infty] = \lim_{z \to 1}\left(1 - z^{-1}\right) X(z) \text{ if the poles of}\left(1 - z^{-1}\right) X(z) \text{ are inside the}$$

unit circle.

In chapter 10, the z-transform will be used to develop the transfer function of difference (or recursive) equations.

## 1.3    SOLVING SIMULTANEOUS EQUATIONS AND SIMPLIFYING ALGEBRAIC EXPRESSIONS

In electrical engineering, one often finds himself finding the solution of two or three simultaneous equations. Here, two methods describe how to solve a system of three equations. The technique also applies to a system of two equations.

Prior to solving a system of simultaneous equations, determine if it has a solution. The following principles are helpful:

- For a solution to exists, the number of equations must exactly equal the number of unknowns. Coefficients of variables in any equation must not be multiple of the corresponding coefficients in another equation.

- If the number of equations is less than the number of unknowns then the system has an infinite number of solutions.

- If the number of equations is greater than number of unknowns, the system has no solution.

### 1.3.1    The method of substitution

Consider a set of three simultaneous equations (with three variables). The method of substitution consists of (1) solving the first variable from the first equation, (2) substituting the variable in the two remaining equations, (3) solving the second variable from the first of the two equations, (4) substituting the variable in remaining single equation, and (5) solving for the third variable. Once the third variable is found, use it to find the second variable. Finally, the two known variables should be used to find the first variable.

Example 1. Solve the following three equations:

$$V_1 + 3V_2 - 2V_3 = 5$$
$$3V_1 + 5V_2 + 6V_3 = 7$$
$$2V_1 + 4V_2 + 3V_3 = 8$$

The first variable is found from the first equation as follows:

$$V_1 = 5 + 2V_3 - 3V_2.$$

Replacing $V_1$ from the last two equations gives
$$-4V_2 + 12V_3 = -8$$
$$-2V_2 + 7V_3 = -2$$

Solving for $V_2$ from the first of the last two equations,

$$V_2 = 2 + 3V_3$$

Substituting $V_2$ on the second of the last two equations gives

$$V_3 = 2$$

In summary, the reduced equations form a diagonal shape

$$V_1 = 5 + 2V_3 - 3V_2$$
$$V_2 = 2 + 3V_3$$
$$V_3 = 2$$

Using the known solution for $V_3$ the other variables may be found as follows:

$$V_3 = 2$$
$$V_2 = 2 + 3V_3 = 8$$
$$V_1 = 5 + 2V_3 - 3V_2 = -15$$

## 1.3.2    Using Cramer's rule

Cramer's rule is a direct method to find the solutions of two or more simultaneous equations. The rule is very efficient in solving relatively small systems such as two or three system of equations. On large systems, it can become very inefficient. Most manual calculations in electrical engineering involve two or three simultaneous equations. It is, therefore, best suited for such applications.

Consider solving the following equations:
$$A_1V_1 + A_2V_2 + A_3V_3 = K_1$$
$$B_1V_1 + B_2V_2 + B_3V_3 = K_2$$
$$C_1V_1 + C_2V_2 + C_3V_3 = K_3$$

The first step in finding the solutions requires writing the coefficient matrix

$$\begin{bmatrix} A_1 & A_2 & A_3 \\ B_1 & B_2 & B_3 \\ C_1 & C_2 & C_3 \end{bmatrix}$$

and the constant matrix

$$\begin{bmatrix} K_1 \\ K_2 \\ K_3 \end{bmatrix}.$$

Next, a ratio of the determinant of the coefficient matrix in the numerator and denominator is formed. Finally, to find the solution of variable $V_i$, replace column $i$ of the determinant in the numerator by the elements of the constant matrix.

For the example, $V_1, V_2$, and $V_3$ are given by:

$$V_1 = \frac{\begin{vmatrix} K_1 & A_2 & A_3 \\ K_2 & B_2 & B_3 \\ K_3 & C_2 & C_3 \end{vmatrix}}{\begin{vmatrix} A_1 & A_2 & A_3 \\ B_1 & B_2 & B_3 \\ C_1 & C_2 & C_3 \end{vmatrix}}$$

$$V_2 = \frac{\begin{vmatrix} A_1 & K_1 & A_3 \\ B_1 & K_2 & B_3 \\ C_1 & K_3 & C_3 \end{vmatrix}}{\begin{vmatrix} A_1 & A_2 & A_3 \\ B_1 & B_2 & B_3 \\ C_1 & C_2 & C_3 \end{vmatrix}}$$

$$V_3 = \frac{\begin{vmatrix} A_1 & A_2 & K_1 \\ B_1 & B_2 & K_2 \\ C_1 & C_2 & K_3 \end{vmatrix}}{\begin{vmatrix} A_1 & A_2 & A_3 \\ B_1 & B_2 & B_3 \\ C_1 & C_2 & C_3 \end{vmatrix}}$$

Notice how the constant matrix replaces the corresponding column of the unknown variable in the coefficient matrix.

The determinant of any 3 X 3 matrix may be found using the pattern below:

$$\begin{vmatrix} A_1 & A_2 & A_3 \\ B_1 & B_2 & B_3 \\ C_1 & C_2 & C_3 \end{vmatrix} = A_1 \begin{vmatrix} B_2 & B_3 \\ C_2 & C_3 \end{vmatrix} - A_2 \begin{vmatrix} B_1 & B_3 \\ C_1 & C_3 \end{vmatrix} + A_3 \begin{vmatrix} B_1 & B_2 \\ C_1 & C_2 \end{vmatrix}$$

Each 2 X 2 determinant above may be evaluated by the pattern:

$$\begin{vmatrix} B_2 & B_3 \\ C_2 & C_3 \end{vmatrix} = B_2 C_3 - C_2 B_3$$

Example 2. Using example 1, find $V_1$ from the system of equations. $V_1$ is given by

$$V_1 = \frac{\begin{vmatrix} K_1 & A_2 & A_3 \\ K_2 & B_2 & B_3 \\ K_3 & C_2 & C_3 \end{vmatrix}}{\begin{vmatrix} A_1 & A_2 & A_3 \\ B_1 & B_2 & B_3 \\ C_1 & C_2 & C_3 \end{vmatrix}}$$

Plugging in the numbers,

$$V_1 = \frac{\begin{vmatrix} 5 & 3 & -2 \\ 7 & 5 & 6 \\ 8 & 4 & 3 \end{vmatrix}}{\begin{vmatrix} 1 & 3 & -2 \\ 3 & 5 & 6 \\ 2 & 4 & 3 \end{vmatrix}}.$$

Finally,

$$V_1 = \frac{1(15-24)-3(9-12)-2(12-10)}{5(15-24)-3(21-48)-2(28-40)} = -15.$$

The result is the same as in example 1.

## 1.4    SIMPLIFYING ALGEBRAIC EXPRESSIONS

As the reader may imply, deriving an accurate transfer function of a circuit or a system is the most important step in the analysis and realization of electronic systems. During analysis, the transfer function is in the form of product of factors (during synthesis or realization the transfer function is in the form of sum of terms). Expressions in a transfer function, if they are factors, must be put in a standard form. The standard form provides quick interpretation of circuit parameters such as -3 dB cutoff point, resonant frequency, damping ratio, time constants, and so on. To transform an expression to its standard form it must be simplified.

In the following, $e$ is an expression, $N$ is a numerator, $D$ is a denominator, and $C$ is any constant or a parameter. The following are the four most common forms of expressions in electrical engineering:

1.  $e_1 = \dfrac{N_1}{D_1} + \dfrac{N_2}{D_2} = \dfrac{N_1}{D_1}\left(\dfrac{D_2}{D_2}\right) + \dfrac{N_2}{D_2}\left(\dfrac{D_1}{D_1}\right) = \dfrac{N_1 D_2 + N_2 D_1}{D_1 D_2}$

2.  $e_2 = \dfrac{N_1}{D_1 D_2} = \dfrac{N_1\left(\dfrac{1}{C}\right)}{D_1 D_2\left(\dfrac{1}{C}\right)} = \dfrac{N_1\left(\dfrac{1}{C}\right)}{D_1\left(\dfrac{1}{C}\right) D_2}$

3.  $e_3 = \dfrac{N_1}{D_1 D_2} = \dfrac{N_1\left(\dfrac{1}{C_1}\right)}{D_1 D_2\left(\dfrac{1}{C_1}\right)} = \dfrac{N_1\left(\dfrac{1}{C_1}\right)\left(\dfrac{1}{C_2}\right)}{D_1\left(\dfrac{1}{C_1}\right) D_2\left(\dfrac{1}{C_2}\right)}$

4.  $e_4 = F_1 \cdot S_1 + F_2 \cdot S_2 = F \cdot S$

For the first three expressions above, one simple rule is applied. That is, if the numerator and denominator of a ratio are multiplied or divided by the same number, the ratio is unchanged. The rule is logical, symmetrical, and easy to implement. It does not require a recall of formulas for simplifying an expression.

The first expression is a sum of two ratios. Simply multiply the numerator and denominator of the first ratio by the denominator of the second ratio. Similarly, multiply the numerator and denominator of the second ratio by the denominator of the first ratio. The result is an equivalent expression with common denominator. Next, simply add the new numerators.

Expression number two, with product of two denominators, multiplies the numerator and denominator by $\dfrac{1}{C}$. This is also equivalent to dividing the numerator and denominator by $C$. The expression also shows the commutative property of multiplication. That is, $\dfrac{1}{C}$ may be placed next to any denominator as long as the denominators are factors.

The third expression uses two multipliers to form an equivalent expression. It is an extension of expression number two.

There are cases when an expression consists of a sum of products of factors and sums of terms in parentheses. Expression 4 shows the pattern. To simplify such an expression, use the distributive property of algebra to distribute each factor onto the terms in a sum. Repeat for the other products of factor and sum. Finally, collect the like terms. The result must consist of one factor and sum only.

As mentioned before, expressions in a transfer function must be in a standard form. That form is stated below:

> The highest power of the Laplace variable, $s$, in a denominator  must have a coefficient of 1. Additionally, $s$ must be the first literal of a term.

Examples of forms that are not allowed are: (1) $as + b$, and (2) $ks^2 + ms + n$. The following examples show how to transform such expressions.

Example 3. Transform $\dfrac{R}{sL+R}$ to its standard form.

For $s$ to have a coefficient of 1, its coefficient $L$, must be factored. First, divide the numerator and denominator by $L$. Next, do the actual division of $sL+R$ by $L$. The following is the result:

$$\frac{\dfrac{R}{L}}{\dfrac{sL+R}{L}} = \frac{R/L}{s+R/L}.$$

The expressions in the numerator and denominator are now in standard form.

Example 4. Transform $\dfrac{p}{(as+b)(ks^2+ms+n)}$ to its standard form.

The letter $a$ must be separated from $as$, and $k$ from $ks^2$. Dividing the numerator and the denominator by $ak$ and carrying out the actual division gives,

$$\frac{p/(ak)}{\dfrac{(as+b)}{a}\dfrac{(ks^2+ms+n)}{k}} = \frac{p/(ak)}{\left(s+\dfrac{b}{a}\right)\left(s^2+s\dfrac{m}{k}+\dfrac{n}{k}\right)}.$$

Again, the expressions in the ratio are in the required form. The following example requires two operations.

Example 5. Transform $\dfrac{s}{3}+\dfrac{1}{2}$ to standard form.

The example has two ratios. For the first operation, multiply the numerator and denominator of each ratio by the same number. That number must be the denominator of the other ratio. Simplify the two ratios to a single ratio.

$$\left(\frac{2}{2}\right)\frac{s}{3}+\left(\frac{3}{3}\right)\frac{1}{2}=\frac{2s+3}{6}.$$

For the last operation, separate the coefficient of $s$. That is,

$$\frac{(2s+3)/2}{6/2}=\frac{s+\dfrac{3}{2}}{3}.$$

It is now in standard form.

Example 5. Simplify $s\left(s+\dfrac{1}{s+2}\right)+s\left(\dfrac{1}{3}+\dfrac{1}{s}\right)$.

The given expression follows the pattern of expression 4. Applying the distributive property of algebra,

$$s\left(s+\frac{1}{s+2}\right)+s\left(\frac{1}{3}+\frac{1}{s}\right)=s^2+\frac{s}{s+2}+\frac{s}{3}+1.$$

Simplifying further,

$$=\frac{3(s+2)}{3(s+2)}s^2+\frac{(3)}{(3)}\frac{s}{s+2}+\frac{(s+2)}{(s+2)}\frac{s}{3}+\frac{3(s+2)}{3(s+2)}1.$$

The expression has now a common denominator. Collecting the terms in each numerator to form a single numerator,

$$= \frac{6s^2 + 3s^3 + 3s + 6 + s^2 + 6 + 3s}{3(s+2)}.$$

Collecting like terms in the numerator and making the coefficient of the highest power of $s$ :

$$= \frac{3s^3 + 7s^2 + 6s + 12}{3(s+2)}$$

$$= \frac{\left(\dfrac{1}{3}\right)\left[3s^3 + 7s^2 + 6s + 12\right]}{\left(\dfrac{1}{3}\right)3(s+2)}.$$

Finally, the simplified expression is

$$= \frac{s^3 + \dfrac{7s^2}{3} + 2s + 4}{s+2}.$$

# Chapter 2

# Transfer Function

Fundamental in the analysis and realization of any control system is its transfer function. It is the transfer function that is used in the analyzing the stability of a control system.

The author recommends using node admittance matrix in formulating a transfer function. Node admittance matrix represents, in final and compact form, the equations of a circuit. It minimizes the lengthy derivations of (voltage) loop equations and (current) node equations.

## 2.1      TRANSFER FUNCTION OF BASIC CIRCUIT USING VOLTAGE DIVISION

Figures 2.1 through 2.3 show the transfer function of the basic RL, RC, and RLC circuits. All of them have the variable $s$. Voltage division was used to get each transfer function.

Essentially, the voltage across a load is equal to the source voltage multiplied by the ratio of the load impedance and the total impedance seen by the source. Hence, the ratio of the load and source voltages, or transfer function, is the ratio of the load impedance and the total impedance seen by the source.

The figures show that the denominator of the transfer function of a circuit is the same. For example, independent of whether the voltage is taken across a resistor or inductor, the denominator of the transfer functions of the RL circuit is the same. The same is true for the RC circuit and the RLC

NOTE:

1. THE DENOMINATOR OF THE TRANSFER FUNCTION IN
BOTH CIRCUITS IS THE SAME.

$$\frac{V_{out}}{V_{in}}(s) = \frac{R/L}{s + R/L}$$

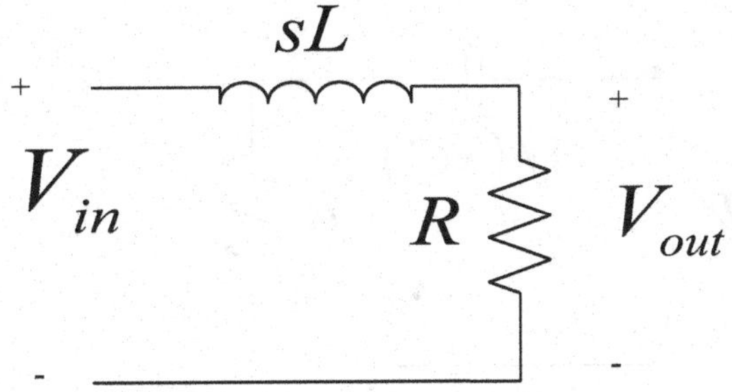

(a) Output taken across the resistor

$$\frac{V_{out}}{V_{in}}(s) = \frac{s}{s + R/L}$$

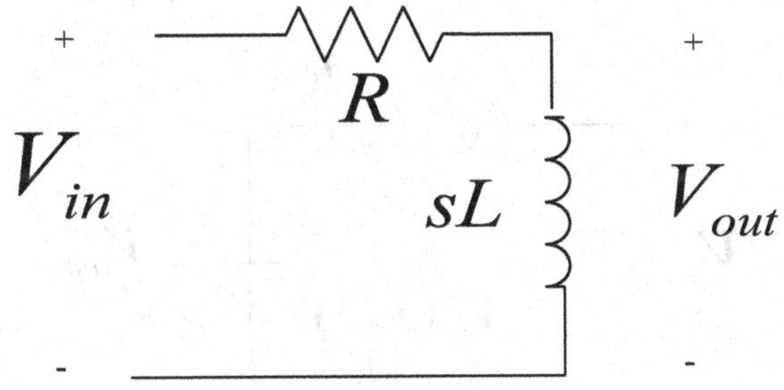

(b) Output taken across the inductor

Figure 2.1 Transfer Function of RL Circuit

NOTE:

1. THE DENOMINATOR OF THE TRANSFER FUNCTION IN
BOTH CIRCUITS IS THE SAME.

$$\frac{V_{out}}{V_{in}}(s) = \frac{s}{s + 1/(RC)}$$

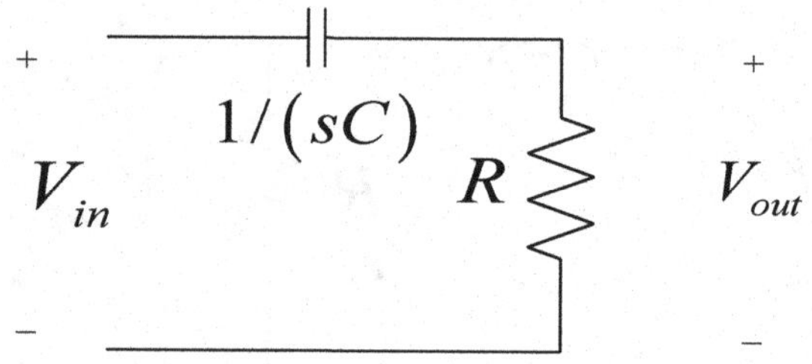

(a) Output taken across the resistor

$$\frac{V_{out}}{V_{in}}(s) = \frac{1/(RC)}{s + 1/(RC)}$$

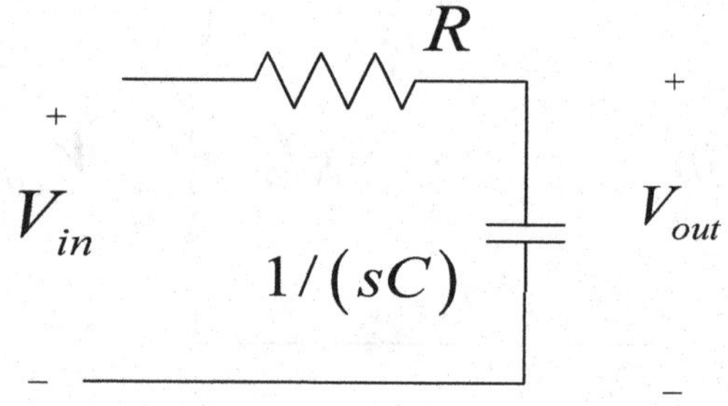

(b) Output taken across the capacitor

Figure 2.2 Transfer Function of RC Circuit

NOTE:

1. THE DENOMINATOR OF THE TRANSFER FUNCTION IN BOTH CIRCUITS IS THE SAME.

$$\frac{V_{out}}{V_{in}}(s) = \frac{s(R/L)}{s^2 + s(R/L) + 1/(LC)}$$

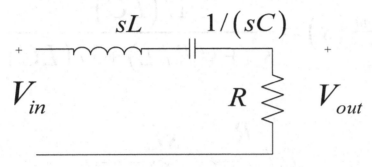

(a) Output taken across the resistor

$$\frac{V_{out}}{V_{in}}(s) = \frac{s(R/L)}{s^2 + s(R/L) + 1/(LC)}$$

(b) Output taken across the inductor

Figure 2.3 Transfer Function of Series RLC Circuit

NOTE:

1. THE DENOMINATOR OF THE TRANSFER FUNCTION IN BOTH CIRCUITS IS THE SAME.

$$\frac{V_{out}}{V_{in}}(s) = \frac{1/(LC)}{s^2 + s(R/L) + 1/(LC)}$$

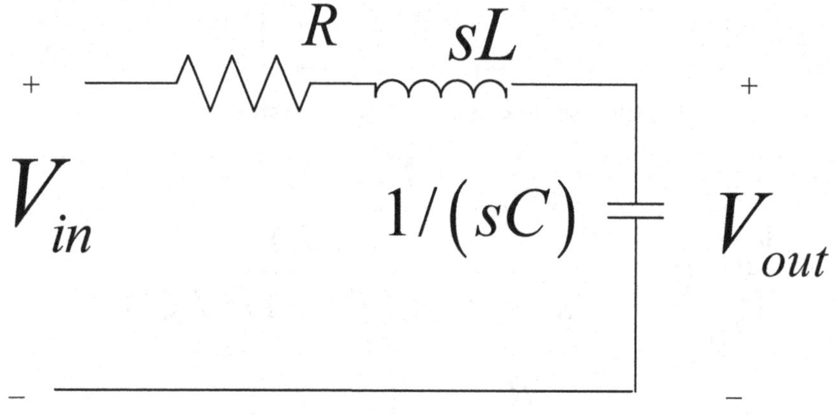

(c) Output taken across the capacitor

Figure 2.3 (continued) Transfer Function of a Series RLC Circuit

circuit. Note, however, that the order of the denominator in the RL or RC circuit is one while the RLC circuit has order of two.

## 2.2　　TRANSFER FUNCTION USING NODE ADMITTANCE MATRIX

The node admittance matrix provides a mechanical way in deriving the voltage across any node. Its basis is Kirchoff's voltage and current laws. However, when applied, getting the node admittance matrix does not require applying the laws. It simply requires entering the correct elements in the source matrix, the admittance matrix, and the node matrix.

## 2.2.1　　Basis of the node admittance matrix

To derive the basis of the approach, see Figure 2.4. The current through $R_1$ is

$$i_1 = \frac{V_1 - V_2}{R_1}.$$

Simplifying,

$$i_1 = \frac{V_1}{R_1} - \frac{V_2}{R_1}$$

or, in matrix form

$$i_1 = \left[ \frac{1}{R_1} - \frac{1}{R_1} \right] \begin{bmatrix} V_1 \\ V_2 \end{bmatrix}.$$

The current through $R_2$ is

$$i_2 = \frac{V_2}{R_2}.$$

INITIAL EQUATIONS USING KIRCHOFF'S
CURRENT LAW:

$$i_1 = \frac{V_1 - V_2}{R_1}$$

$$i_2 = \frac{V_2}{R_2}$$

$$i_1 = i_2$$

$$\frac{V_1 - V_2}{R_1} = \frac{V_2}{R_2}$$

FINAL RESULT IN MATRIX
FORM:

$$\begin{bmatrix} i_1 \\ 0 \end{bmatrix} = \begin{bmatrix} \dfrac{1}{R_1} & -\dfrac{1}{R_1} \\ -\dfrac{1}{R_1} & \left( \dfrac{1}{R_1} + \dfrac{1}{R_2} \right) \end{bmatrix} \begin{bmatrix} V_1 \\ V_2 \end{bmatrix}$$

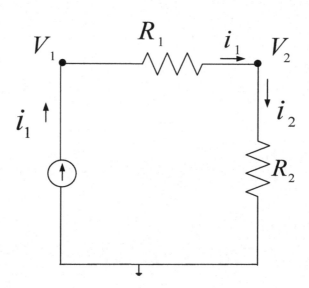

Figure 2.4 Sample Circuit for Deriving Node Admittance Matrix

Since $i_1 = i_2$,

$$\frac{V_1 - V_2}{R_1} = \frac{V_2}{R_2}.$$

Simplifying,

$$0 = -\frac{V_1}{R_1} + \left(\frac{1}{R_1} + \frac{1}{R_2}\right) V_2$$

or, in matrix form

$$0 = \left[-\frac{1}{R_1} + \left(\frac{1}{R_1} + \frac{1}{R_2}\right)\right]\begin{bmatrix} V_1 \\ V_2 \end{bmatrix}.$$

Both matrices may be combined into a single matrix equation:

$$\begin{bmatrix} i_1 \\ 0 \end{bmatrix} = \begin{bmatrix} \dfrac{1}{R_1} & -\dfrac{1}{R_1} \\ -\dfrac{1}{R_1} & \left(\dfrac{1}{R_1} + \dfrac{1}{R_2}\right) \end{bmatrix}\begin{bmatrix} V_1 \\ V_2 \end{bmatrix}.$$

The matrix equation above has the current matrix (left hand side), the admittance matrix, and the node voltage matrix.

Focus on the entries of the admittance matrix. First, the elements in the diagonal of the matrix are the sum of the all admittances connected to a node. For example, the sum of the admittances in node 1 is $\dfrac{1}{R_1}$. Node 2, however, has two admittances connected to it. It is $\left( \dfrac{1}{R_1} + \dfrac{1}{R_2} \right)$. Now entries in the off-diagonal elements are the negative of the sum of the admittance between the two nodes. For example, between nodes 1 and 2, the sum of the admittance is $-\dfrac{1}{R_1}$. Between nodes 2 and 1 the entry must also be $-\dfrac{1}{R_1}$.

Admittance matrix is symmetrical around its diagonal elements. Thus, the entry in row M and column N must be the same as the entry in row N and column M.

The efficiency of the node admittance matrix lies on the simplicity of the above rules. Note that no circuit analysis was involved in determining an entry. Node admittance matrix may be extended to any number of nodes.

## 2.2.2 Deriving the transfer function using the node admittance matrix

A transfer function is always defined relative to the voltage source and not the current source. The key is to transform the current source into a voltage source. If a circuit has voltage source, $V_S$, in series with source resistance, $R_S$, then the current source is

$$i_1 = \frac{V_S}{R_S}.$$

Thus, to get a transfer function of a circuit $\dfrac{V_S}{R_S}$ must be used for the current source. In addition, the admittance of the source resistance must be considered when finding the admittance connected to node 1. The resulting

INITIAL EQUATIONS USING
KIRCHOFF'S CURRENT LAW:

$$i_1 = \frac{V_S}{R_S} = \frac{V_1 - V_2}{R_1}$$

$$i_2 = \frac{V_2}{R_2}$$

$$i_1 = i_2$$

$$\frac{V_1 - V_2}{R_1} = \frac{V_2}{R_2}$$

RESULT IN MATRIX FORM:

$$\begin{bmatrix} \dfrac{V_S}{R_S} \\ 0 \end{bmatrix} = \begin{bmatrix} \dfrac{1}{R_1} + \dfrac{1}{R_S} & -\dfrac{1}{R_1} \\ -\dfrac{1}{R_1} & \left(\dfrac{1}{R_1} + \dfrac{1}{R_2}\right) \end{bmatrix} \begin{bmatrix} V_1 \\ V_2 \end{bmatrix}$$

TRANFER FUNCTION:

$$\frac{V_2}{V_S} = \frac{R_2}{R_1 + R_2 + R_S}$$

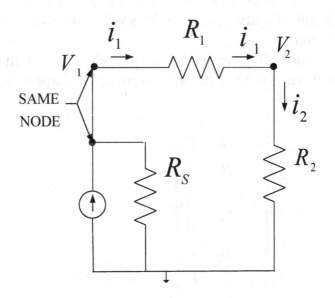

Figure 2.5 The Node Admittance Matrix of a more Realistic Circuit

circuit, matrices, and transfer function of $\dfrac{V_2}{V_s}$ are shown on Figure 2.5.

Using voltage division shows that the transfer function is correct.

## 2.2.3 Transfer function of a more complicated circuit

Figure 2.6 shows a circuit with one capacitor and two inductors. It has four nodes (excluding the reference or ground node). However, the interest is in getting the transfer function $\dfrac{V_3}{V}$. Hence, the node between the second resistor and the first inductor will be merged into a single node. This will reduce the number of equations from four to three. Figure 2.7 shows the admittance representation of the elements in the circuit.

In addition to showing the merging of nodes, Cramer's rule and the rules on simplifying expressions will be shown in deriving the transfer function.

The voltage across the third node of a system of three simultaneous equations is given by Cramer's rule (see chapter 1),

$$V_3 = \frac{\begin{vmatrix} A_1 & A_2 & K_1 \\ B_1 & B_2 & K_2 \\ C_1 & C_2 & K_3 \end{vmatrix}}{\begin{vmatrix} A_1 & A_2 & A_3 \\ B_1 & B_2 & B_3 \\ C_1 & C_2 & C_3 \end{vmatrix}}$$

Using the node admittance matrix,

$$R_1 = 1 \qquad C = 1$$
$$R_2 = 2 \qquad L_1 = 1$$
$$R_3 = 3 \qquad L_2 = 2$$

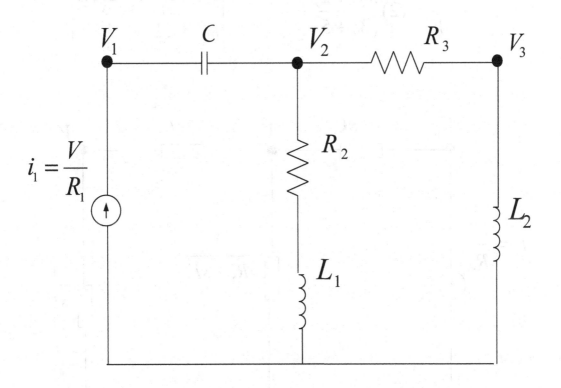

Figure 2.6 A Relatively Complex Circuit

PARAMETERS:

$R_1 = 1 \qquad C = 1$

$R_2 = 2 \qquad L_1 = 1$

$R_3 = 3 \qquad L_2 = 2$

TRANSFER FUNCTION:

$$\frac{V_3}{V} = (2)\frac{s(s+2)}{(3s+5)}$$

VOLTAGE ACROSS NODE 3:

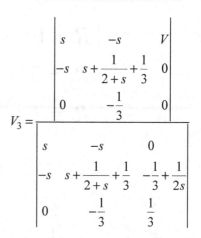

$$V_3 = \frac{\begin{vmatrix} s & -s & V \\ -s & s + \dfrac{1}{2+s} + \dfrac{1}{3} & 0 \\ 0 & -\dfrac{1}{3} & 0 \end{vmatrix}}{\begin{vmatrix} s & -s & 0 \\ -s & s + \dfrac{1}{2+s} + \dfrac{1}{3} & -\dfrac{1}{3} + \dfrac{1}{2s} \\ 0 & -\dfrac{1}{3} & \dfrac{1}{3} \end{vmatrix}}$$

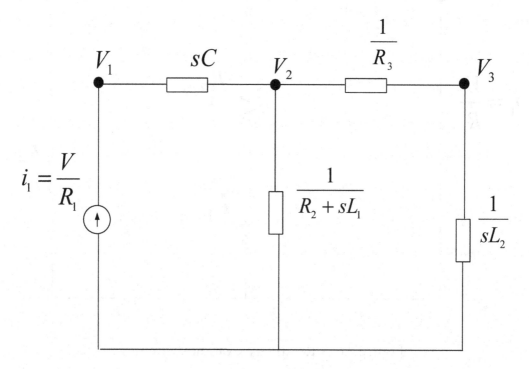

Figure 2.7 Admittances of Circuit Elements

$$V_3 = \frac{\begin{vmatrix} sC & -sC & \dfrac{V}{R_1} \\[3mm] -sC & sC + \dfrac{1}{R_2+sL_1} + \dfrac{1}{R_3} & 0 \\[3mm] 0 & -\dfrac{1}{R_3} & 0 \end{vmatrix}}{\begin{vmatrix} sC & -sC & 0 \\[3mm] -sC & sC + \dfrac{1}{R_2+sL_1} + \dfrac{1}{R_3} & -\dfrac{1}{R_3} \\[3mm] 0 & -\dfrac{1}{R_3} & \dfrac{1}{R_3} + \dfrac{1}{sL_2} \end{vmatrix}} .$$

The presence of some zeros in the determinants above will greatly reduce the effort in simplifying expressions. Such zeros make the matrices of circuits sparse (most entries are zero). Also, the matrix is symmetrical relative to its diagonal elements. That is, an element on top of the diagonal must also appear on the bottom of the diagonal. This is a self-checking feature of the node admittance matrix.

Substituting the parameter values,

$$V_3 = \frac{\begin{vmatrix} s & -s & V \\ -s & s + \dfrac{1}{2+s} + \dfrac{1}{3} & 0 \\ 0 & -\dfrac{1}{3} & 0 \end{vmatrix}}{\begin{vmatrix} s & -s & 0 \\ -s & s + \dfrac{1}{2+s} + \dfrac{1}{3} & -\dfrac{1}{3} + \dfrac{1}{2s} \\ 0 & -\dfrac{1}{3} & \dfrac{1}{3} \end{vmatrix}}$$

To simplify the above equation, represent the determinant in the numerator as $N$:

$$N = \frac{sV}{3}.$$

Similarly, represent the determinant in the denominators as $D$:

$$D = s\left[\left(s + \frac{1}{2+s} + \frac{1}{3}\right)\left(\frac{1}{3} + \frac{1}{2s}\right) - \frac{1}{9}\right] - (-s)\left[(-s)\left(\frac{1}{3} + \frac{1}{2s}\right)\right].$$

Next, apply the recommended rules for simplifying expressions. The following is the sequence of such simplification:

$$D = \frac{s^2}{3} + \frac{s}{3(s+2)} + \frac{s}{2} + \frac{1}{2(s+2)} + \frac{1}{6} - \frac{s^2}{3} - \frac{s}{2}$$

$$= \frac{s}{3(s+2)} + \frac{1}{2(s+2)} + \frac{1}{6}$$

$$= \frac{(2)}{(2)} \frac{s}{3(s+2)} + \frac{(3)}{(3)} \frac{1}{2(s+2)} + \frac{(s+2)}{(s+2)} \frac{1}{6}$$

$$= \frac{2s + 3 + s + 2}{6(s+2)}$$

$$= \frac{3s + 5}{6(s+2)}.$$

The denominator is now in the standard form. Simplifying for $V_3$,

$$V_3 = \frac{N}{D}$$

$$= \frac{\dfrac{sV}{3}}{\dfrac{3s+5}{6(s+2)}}$$

$$= \frac{sV}{3} \frac{6(s+2)}{(3s+5)}.$$

Dividing both sides by $V$ gives

$$\frac{V_3}{V} = (2)\frac{s(s+2)}{(3s+5)}.$$

Simplifying further the transfer function is

$$\frac{V_3}{V} = \frac{2s^2 + 4s}{3s + 5}.$$

At this point, the interest is simply on using the admittance matrix in deriving the transfer function of a circuit. Rigorous interpretation of the meaning of the function will be shown in chapters 6 through 8. Specifically, chapter 8 will show how a transfer function is used in the realization or design of the control system the function represents.

# Chapter 3

# Block Diagrams

A block diagram consists of boxes representing the transfer function of subsystems in a system. Subsystems may be an electrical circuit, actuator, transducer, or any other device. Boxes in a block diagram must be reduced to a single block. The single block represents the transfer function of the whole system. In addition to showing some block reduction rules, this chapter also shows how a transfer may be derived from a block diagram.

## 3.1     A SIMPLE BLOCK DIAGRAM

A block diagram depicts how subsystems are integrated together to form a system. In its basic form, a block diagram can be represented as shown on Figure 3.1. The figure shows an input $R$, a block $G$, and output $C$. $R$, $G$, and $C$ are all function of the Laplace variable s. They are related by

$$C = GR.$$

The above equation implements the convolution theorem (in the s domain). When two blocks are in cascade, then their equivalent block is the product of the two blocks.

Figure 3.1 shows a block diagram with no feedback. When a feedback exists, a new formula exists between the output and the input.

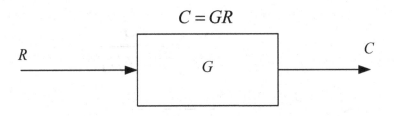

$$C = GR$$

Figure 3.1 A Simple Block Diagram

## 3.2 BLOCK DIAGRAM WITH FEEDBACK

Figure 3.2 is a block diagram with feedback. The feed forward block is $G$. Its feed backward block is $H$. A comparator (or summer) compares the input signal $R$ and the feedback signal $B$. Note that the input has a positive sign and the feedback has a negative sign. The difference between the two signals is designated as the error, $e$.

The feedback signal is the product of the output signal and the feed backward block.

$$B = CH .$$

In contrast the error, $e$, is the difference between the input and the feedback signal.

$$e = R - B .$$

Substituting for $B$ gives

$$e = R - CH .$$

$$\frac{C}{R} = \frac{G}{1+GH}$$

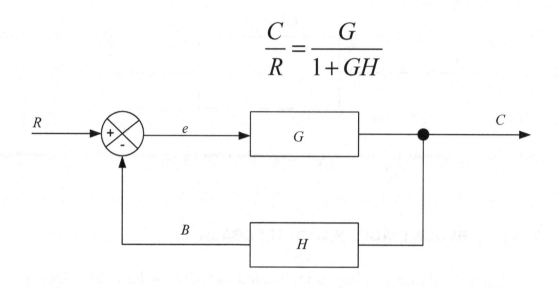

Figure 3.2 Block Diagram with Forward and Feedback Paths

Now, express the output as the product of the error signal and the feed forward block

$$C = eG.$$

Substituting $e = R - CH$ gives

$$C = (R - CH)G.$$

Expanding the above, collecting like terms, and factoring gives

$$C = RG - CGH,$$

and

$$C(1+GH) = RG.$$

Finally

$$\frac{C}{R} = \frac{G}{1+GH}.$$

Note that when the denominator, $1 + GH$, is equal to zero then $C/R$ becomes undefined. This is the unstable case. Much of the techniques developed in the analysis and synthesis of electronic circuits involves stabilizing $C/R$.

## 3.3     OTHER RULES IN REDUCING A BLOCK DIAGRAM

Figure 3.3(a) is an example of two blocks in parallel. Its equivalent is the sum of the blocks. This is shown on Figure 3.3(b).

There are cases when the pickoff point of a feedback signal is not the final output. For example in Figure 3.4(a), the pickoff point is indicated by point 1. To move the pickoff point from 1 to 2, the block in the feedback path must be changed. The steps are shown on Figure 3.4(b) and 3.4(c). Figure 3.4(c) also shows the derivation of the resulting equivalent block. Essentially, the equivalent block in the feedback path is divided by the block between the initial pickoff point and the final pickoff point.

## 3.4     SIMPLIFYING A COMPLEX BLOCK DIAGRAM

Consider simplifying a relatively complex block diagram as shown on Figure 3.5. The block reduction rules described above may be used to

*Derivation:*

$$h = CH$$
$$g = CG$$
$$h + g = C(H + G)$$

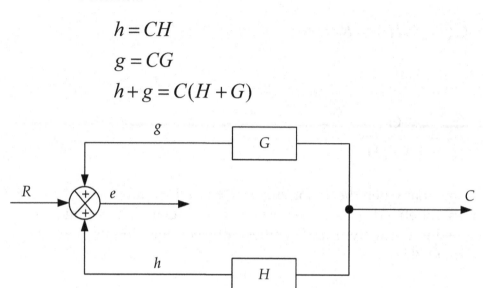

(a) Block Diagram with Two Parallel Feedbacks

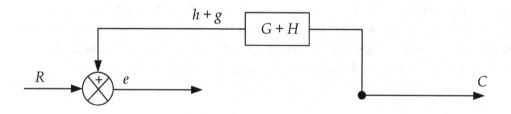

(b) Equivalent Block of the Two Parallel Blocks

Figure 3.3 The Equivalent Block of Two Parallel Feedback Blocks

Derivation:

$$h = \frac{C}{G}$$

$$y \text{ (original)} = hH$$

$$y \text{ (original)} = \left(\frac{C}{G}\right)H$$

$$y \text{ (with X)} = CX$$

$$y \text{ (original)} = y \text{ (with X)}$$

$$X = \frac{H}{G}$$

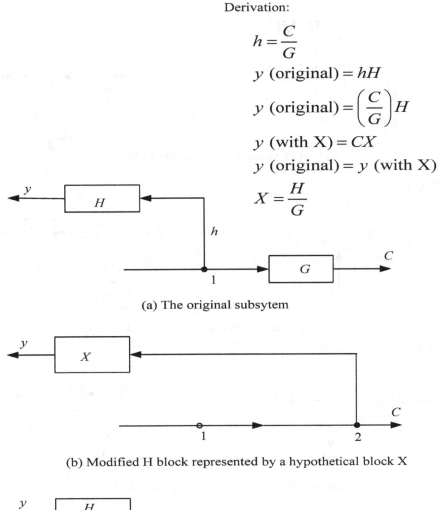

(a) The original subsytem

(b) Modified H block represented by a hypothetical block X

(c) Representation of the X block by H/G block

Figure 3.4 Reducing a Forward Block to Feedback Block

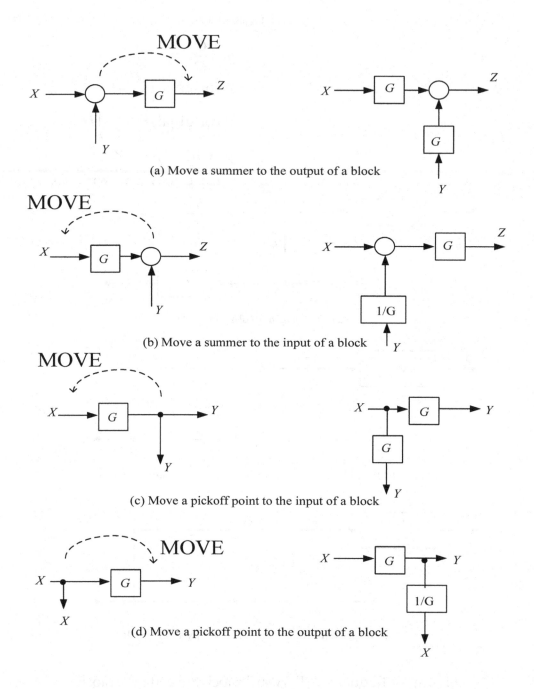

(a) Move a summer to the output of a block

(b) Move a summer to the input of a block

(c) Move a pickoff point to the input of a block

(d) Move a pickoff point to the output of a block

Figure 3.5 Other Block Reduction Rules

simplify the diagram. These rules are shown on Figures 3.6 through 3.10. Note that on Figure 3.9, the resulting feedback signal of the two combined parallel blocks must be negative. Hence, the sign of the sum of the positive feedback and the negative feedback must be negative.

On Figure 3.10, an intermediate input, Y, is placed before the comparator. It will be used to calculate the intermediate transfer function since between Y and R is another block G1. Figure 3.11 is the final simplified diagram.

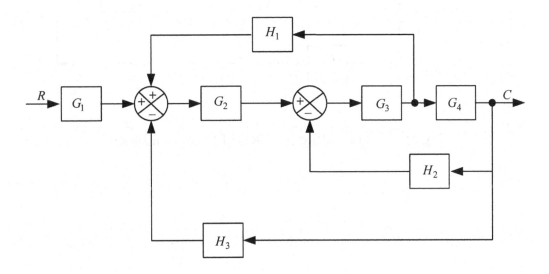

Figure 3.6 Example of a Block Diagram to be Simplified

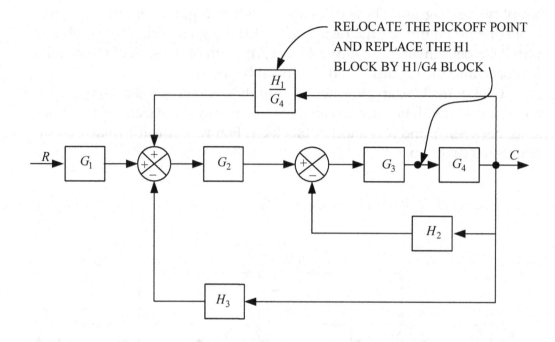

Figure 3.7 Relocating the Pickoff Point of a Block

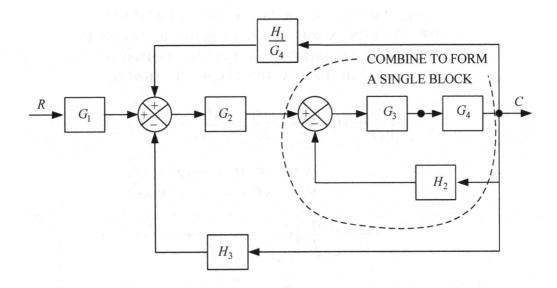

Figure 3.8 Combining Blocks that has a Summer and a Loop of
Forward and Feedback Blocks

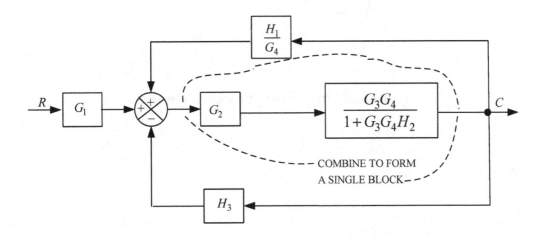

Figure 3.9 Simplifying Blocks in Series

NOTE: MAINTAIN THE NEGATIVE SIGN AT THE SUMMER OF
THE RESULTING SINGLE FEEDBACK BLOCK BY TAKING THE
OPPOSITE OF THE DIFFERENCE BETWEEN THE POSITIVE
FEEDBACK AND THE NEGATIVE FEEDBACK.  THAT IS,

$$-\left(\frac{H_1}{G_4}-H_3\right)=H_3-\frac{H_1}{G_4}=\frac{G_4H_3-H_1}{G_4}$$

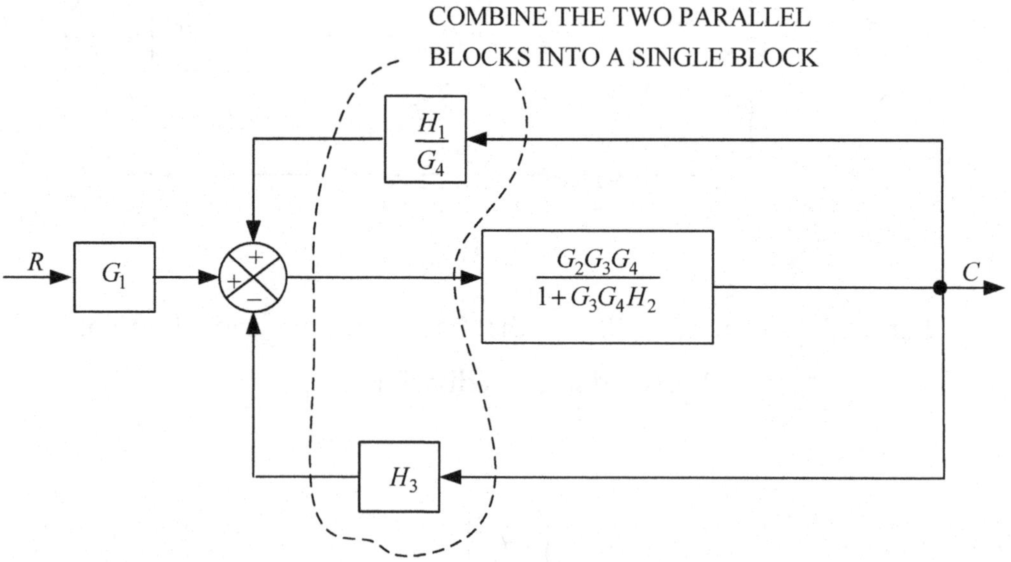

Figure 3.10 Simplifying Blocks in Parallel

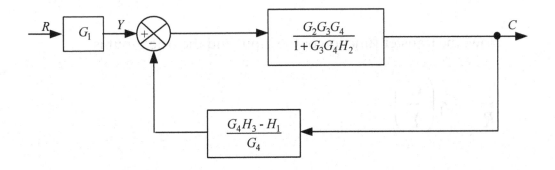

Figure 3.11 The Simplified Block Diagram

## 3.5 TRANSFER FUNCTION OF THE SIMPLIFIED BLOCK DIAGRAM

The ratio of the output and the intermediate input, $Y$, to the comparator is,

$$\frac{C}{Y} = \frac{\dfrac{G_2 G_3 G_4}{1 + G_3 G_4 H_2}}{1 + \dfrac{G_2 G_3 G_4}{(1 + G_3 G_4 H_2)} \dfrac{(G_4 H_3 - H_1)}{G_4}}.$$

Simplifying further,

$$\frac{C}{Y} = \frac{G_2 G_3 G_4}{1 + G_3 G_4 H_2 + G_2 G_3 G_4 H_3 - G_2 G_3 H_1}.$$

Now, the transfer function of the input and the true input is

$$\frac{C}{R} = G_1\left(\frac{C}{Y}\right).$$

## 3.6   INTERPRETATION OF THE TRANSFER FUNCTION

The transfer function above shows that its numerator consists of forward blocks. In contrast, the denominator consists of terms that are the product of forward blocks and feedback blocks. Additionally, the denominator also has also a term with negative sign.

Other important, yet subtle, patterns include:

1. the numerator is the product of all the forward blocks,
2. the denominator is the sum of the  products of a feedback block and the forward blocks encompassed by the feedback block (from its pickoff point to its comparator), and
3. a term in the denominator will have a negative sign if the sign of the feedback path at its comparator is positive (a feedback path by definition must be negative).

The above observation provides the capability to develop a computer algorithm to synthesize the transfer function of a block diagram.

Consider a block diagram with c comparators, p pickoff points, h feedback blocks, and g forward blocks. For each feedback block, associate a terminal point at one of the c comparators. Similarly, assign its initial point at one of the p pickoff points. Do the same for blocks in the forward paths except three coordinates will be required. The first two identifies a pickoff point and a comparator. Its third coordinate is the index of the block. The rest of the algorithm consists of forming products for the numerator and terms for the denominator.

Finally,

$$\frac{C}{R} = \frac{G_1 G_2 G_3 G_4}{1 + G_3 G_4 H_2 + G_2 G_3 G_4 H_3 - G_2 G_3 H_1}.$$

A general proof follows. Consider any complicated block diagram. Assume the worst case when the diagram has sub loops that may be interacting. Then by block reduction rules the diagram may be made simple (removing the interaction). The result should be a block diagram with a single forward block and a single feedback block.

## 3.6.1 Example of a relatively complex block diagram

Figure 3.12 is a relatively complex diagram. Each block shows its gain as a function of $s$.

If block reduction rules are applied and using the actual gains of each block, extensive algebraic manipulation will be required in each step. To minimize the number of such manipulations, represent each block by $G$ if it is in a forward path and by $H$ if in the feedback path. This is shown on Figure 3.13. The approach not only provides a strategic view of reduction but also shows a recursive relation between two steps. This recursive relation proves that any block diagram may computed by inspection.

After representing each block by $G$ and $H$, define the value of each subscripted letter as follows:

Gain of each block:

$$G_1 = 20 \qquad G_2 = \frac{1}{s+4} \qquad G_3 = \frac{1}{s+5} \qquad G_4 = \frac{1}{s}$$

$$H_1 = 1 \quad H_{2,equiv} = 5s \quad H_3 = 1$$

ABBREVIATIONS:

s = LAPLACE VARIABLE

L1, L2, L3 = LOOPS WITH FORWARD AND FEEDBACK BLOCKS

F1, F2, F3 = EQUIVALENT BLOCK OF A BLOCK AND THE LOOP
AFTER THE BLOCK

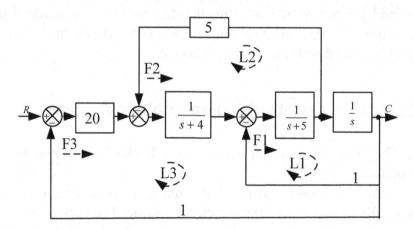

NOTE: WHEN THE FEEDBACK SIGNAL IS EQUAL TO THE OUTPUT
SIGNAL IT IS CUSTOMARY TO REPLACE ITS BLOCK BY 1.

Figure 3.12 A Relatively Complex Block Diagram

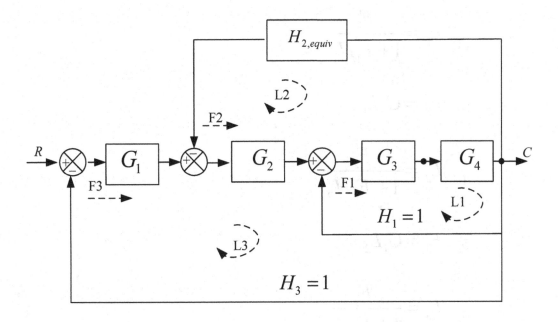

Figure 3.13 A more Abstract Representation of Figure 12 using G's and H's

Note that a block reduction rule was applied in getting the equivalent gain of $H_2$ block.

The next step is to find the transfer function by iteratively simplifying a loop of blocks or a cascade of blocks. Results follow:

1. $F_1 = G_3 G_4$

2. $L_1 = \dfrac{F_1}{1 + F_1 H_1}$

3. $F_2 = G_2 L_1$

4. $L_2 = \dfrac{F_2}{1 + F_2 H_2}$

5. $F_3 = G_1 L_2$

6. $L_3 = \dfrac{F_3}{1 + F_3 H_3}$

7. $F_1 = \dfrac{1}{s^2 + 5s}$

8. $L_1 = \dfrac{1}{s^2 + 5s + 1}$

9. $F_2 = \dfrac{1}{s^3 + 9s^2 + 21s + 4}$

10. $L_2 = \dfrac{1}{s^3 + 9s^2 + 26s + 4}$

11. $F_3 = \dfrac{20}{s^3 + 9s^2 + 26s + 4}$

12. $L_3 = \dfrac{C}{R} = \dfrac{20}{s^3 + 9s^2 + 26s + 26}$.

The transfer function being sought is $L_3$. Notice the extensive number of steps required to get the transfer function. Additionally, each step requires algebraic manipulations to simplify ratios involving $s$. It will be desirable to use a simpler direct approach as shown in the following section.

## 3.6.2    Deriving the transfer function by inspection

The transfer function of Figure 3.12 may also be obtained by simply multiplying all the forward blocks divided by (1 + product of feedback block and all forward blocks encompassed by the feedback path). An example of its form (not related to Figure 3.12) is

$$\frac{C}{R} = \frac{G_1 G_2 G_3 G_4}{1 + G_3 G_4 H_2 + G_2 G_3 G_4 H_3 - G_2 G_3 H_1}$$

## 3.7    BLOCK DIAGRAM OF A DC MOTOR

Industrial control system problems oftentimes involve no more than three loops. As an example, Figure 3.14 is the electrical equivalent of a DC motor and its block diagram. The block diagram shows a single loop only.

The definition of variables, derivation of block gains, and the transfer function is also shown on the figure.

DEFINITION OF VARIABLES:

$s$ = Laplace variable,

$T_P$ = torque produced,

$T_C$ = torque consumed,

$i_f$ = current in the field winding,

$i_a$ = current in the armature winding,

$V$ = applied DC voltage,

$E_{bemf}$ = back electromagnetic force (voltage),

$\theta$ = angular displacement,

$R_a$ = armature resistance,

$R_f$ = field resistance,

$J$ = polar moment of inertia, and

$B$ = moment of inertia.

DERIVATION OF TORQUE, CURRENT, AND BACK EMF:

$$T_P = k_f i_f = k_t i_a$$

$$i_a = \frac{V - E_{bemf}}{R_a}$$

$$E_{bemf} = k_\omega s \theta$$

$$T_C = Js^2 \theta + Bs\theta$$

TRANSFER FUNCTION:

$$\frac{\theta}{V} = \frac{k_t / JR_a}{s^2 + s\left(\dfrac{R_a B + k_t k_\omega}{JR_a}\right)}$$

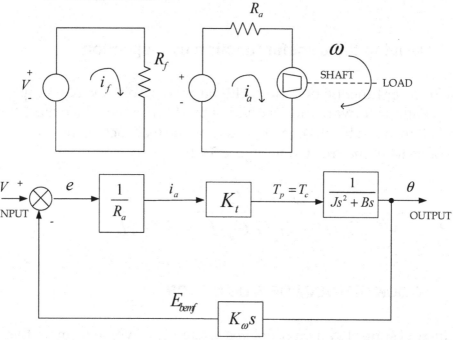

Figure 3.14 The Electrical Equivalent of a DC Motor and its Block Diagram

Perhaps, the derivation of the gains of the blocks is the most challenging part of the model. It involves finding relationships between the torque produced and current, current and back emf, the torque consumed by the shaft and its angular displacement, and finally the angular displacement and the back emf. The back emf is the feedback signal.

In real applications, one performs experiments to discover such relationships. Fortunately, most of the relationships are equations of straight line. For example, the torque produced,

$$T_P = k_f i_f = k_t i_a$$

shows a straight-line relationship between the torque and current in the field winding or armature winding. The constants $k_f$ and $k_t$ are constants representing the slopes of the lines.

# Chapter 4

# Analyzing Signal Flow Graph

Unlike boxes of a block diagram, a signal flow graph shows directed lines representing the paths, gain of each path, and elements of a transfer function. This chapter shows how a system of differential equations can be represented as a signal flow graph. It also describes Mason's rule. The rule is used in deriving the transfer function of a signal flow graph.

## 4.1    SIGNAL FLOW GRAPH

In electrical circuits, the time derivative of current through an inductor and the time derivative of voltage across a capacitor are called state variables. For now, these variables will be represented as $x$.

Consider the following three differential equations:

$$\dot{x}_1 = x_2$$

$$\dot{x}_2 = x_3$$

$$\dot{x}_3 = -6x_1 - 11x_2 - 6x_3 + U \, .$$

Define the Laplace operator, $L$, as an operator that simplifies an expression to its simpler form. The simpler form may be found in a table of Laplace transforms. For example, taking the Laplace transform of the first differential equation gives:

$$L(\dot{x}_1) = L(x_2).$$

Using a table of Laplace transforms,

$$sx_1 = x_2$$

or,

$$x_1 = \frac{x_2}{s} = \left(\frac{1}{s}\right)x_2.$$

The last result links $x1$ with $x2$ by $(1/s)$. This linkage is the basis of signal flow graph.


## 4.2    PROCEDURE FOR DRAWING THE SIGNAL FLOW GRAPH

After formulating the differential equations, the signal flow graph may be drawn as follows:

1. Draw nodes corresponding to twice the number of state variables (see Figure 4.1).
2. Starting from the right, label each node with the variable and its time derivative. Connect both nodes by forward arrow. Label the arrow by $1/s$.
3. Inspect the differential equations and enclose, by an ellipse, two nodes that are equal.
4. Redraw the signal flow graph by replacing each ellipse by the equal nodes (see Figure 4.2).
5. Draw a feedback arrow from a node to another node if a term on the right hand side of the differential equation has a negative sign.

After the signal flow graph is drawn, Mason's rule could be used to derive its transfer function. The rule is given below:

Set of simultaneous differential equations to
be represented by the signal flow graph:

$$\dot{x}_1 = x_2$$

$$\dot{x}_2 = x_3$$

$$\dot{x}_3 = -6x_1 - 11x_2 - 6x_3 + U$$

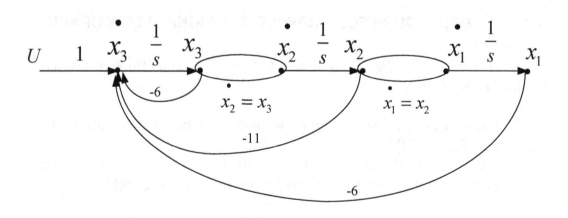

Figure 4.1 A Simple Signal Flow Graph

Set of simultaneous differential equations to be represented by the signal flow graph (each variable is a function of time):

$$\dot{x}_1 = x_2$$

$$\dot{x}_2 = x_3$$

$$\dot{x}_3 = -6x_1 - 11x_2 - 6x_3 + U$$

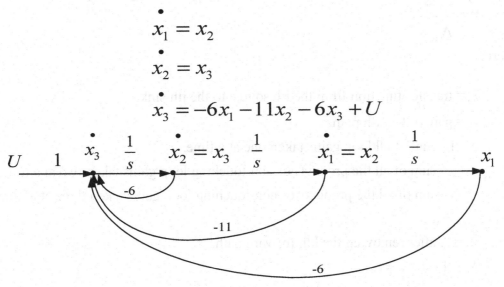

Using the Laplace operator, L, and the Laplace variable s:

$$L(\dot{x}_1) = L(x_2)$$

$$sx_1 = x_2$$

$$x_1 = \frac{x_2}{s} = \left(\frac{1}{s}\right)x_2$$

Transfer function (using Mason's rule):

$$\frac{x_1}{U} = \frac{\dfrac{1}{s^3}}{1 + \dfrac{6}{s} + \dfrac{11}{s^2} + \dfrac{6}{s^3}}$$

$$\frac{x_1}{U} = \frac{1}{s^3 + 6s^2 + 11s + 6}$$

$$\frac{x_1}{U} = \frac{1}{(s+1)(s+2)(s+3)}$$

Figure 4.2 A More Compact Representation of the Signal Flow Graph Shown on Figure 4.1

$$T_{ij} = \frac{\sum P_k \Delta_k}{\Delta}$$

where,

$T_{ij}$ = transfer function from the ith source to the jth sink,

$P_k$ = gain of the kth path,

$\Delta$ = 1 - sum of all loop gains taken one at a time,

+ sum of all the products of non-touching loop gains taken two at a time,

- sum of all the products of non-touching loop gains taken three at a time,

+ ...

$\Delta_k$ = $\Delta$ after removing the kth forward path.

$\Delta_k = 1 + 0 + ... = 1$.

Note that the "sum of all loop gains taken one at a time" is independent of whether the loops are touching or not. However, the "sum of all the products of non-touching loop gains taken two at a time" and the remaining terms require that the loops be non-touching. Non-touching means the loop could not touch either a node or edge of another loop.

The factor $\Delta_k$ must also be clarified. When a forward path touches all loops and the path is removed, then there is no more loop remaining. Hence,

$$\Delta_k = 1 + 0 + ... = 1.$$

A signal flow graph may also have several paths. It does not follow that a forward path can only be forward. There are paths that may return backward and then continue forward. The most important criterion in defining a path is as follows:

A path may return but it must not traverse any loop.

Because of the above constraints, Mason's rule must be implemented with care. It is robust approach but can be error prone. Mason's rule does provide helpful insights on the mechanisms of a transfer function.

To illustrate the rule, examine Figure 4.2. It has the following loop gains:

$$L_1 = -\frac{6}{s} \qquad \text{(touches all)}$$

$$L_2 = -\frac{11}{s^2} \qquad \text{(touches all)}$$

$$L_3 = \frac{-6}{s^3} \qquad \text{(touches all)}.$$

There is only one forward path,

$$P_1 = \frac{1}{s^3} \quad \text{(touches all)}$$

and

$$\Delta_1 = 1 \quad \text{(forward path touches all)}$$

The transfer function is therefore,

$$\frac{x_1}{U} = \frac{\dfrac{1}{s^3}}{1 + \dfrac{6}{s} + \dfrac{11}{s^2} + \dfrac{6}{s^3}}$$

Simplifying further,

$$\frac{x_1}{U} = \frac{1}{(s+1)(s+2)(s+3)}$$

## 4.2.1    Example of a more complicated signal flow graph

Figure 4.3 is a more complicated signal flow graph. While it represents a set of differential equations, such a set will not be shown. Its purpose is to show the intricacies of Mason rule in deriving a transfer function.

Examination of Figure 4.3 shows the following loops in the signal flow graph:

$$L_1 = \frac{-3}{s} \qquad \text{does not touch } L_2, L_3$$

$$L_2 = \frac{-2}{s} \qquad \text{does not touch } L_1, L_3, L_6$$

$$L_3 = \frac{-1}{s} \qquad \text{does not touch } L_1, L_2, L_4$$

$$L_4 = \frac{-4}{s^2} \qquad \text{does not touch } L_3$$

LOOP GAINS:

$L_1 = \dfrac{-3}{s}$    does not touch $L_2, L_3$

$L_2 = \dfrac{-2}{s}$    does not touch $L_1, L_3, L_6$

$L_3 = \dfrac{-1}{s}$    does not touch $L_1, L_2, L_4$

$L_4 = \dfrac{-4}{s^2}$    does not touch $L_3$

$L_5 = \dfrac{8}{s^3}$    touches all

$L_6 = \dfrac{-4}{s^2}$    does not touch $L_2$

NON-TOUCHING LOOPS
TAKEN TWO AT A TIME:

$L_1 L_2$

$L_1 L_3$

$L_2 L_3$

$L_2 L_6$

$L_3 L_4$

NON-TOUCHING LOOPS
TAKEN THREE AT A TIME:

$L_1 L_2 L_3$

PATHS

$P_1 = \dfrac{2}{s^3}$    touches all

$P_2 = \dfrac{1}{s^2}$    does not touch $L_2$

$P_3 = \dfrac{1}{s^2}$    does not touch $L_3$

$P_4 = -\dfrac{2}{s}$    touches all

$$\sum P_k \Delta_k = \frac{2}{s^3} + \frac{1}{s^2}\left(1 + \frac{2}{s}\right) + \frac{1}{s^2}\left(1 + \frac{3}{s}\right) - \frac{2}{s^3}$$

$$\Delta = 1 + \left(\frac{6}{s} + \frac{8}{s^2} + \frac{8}{s^3}\right) + \left(\frac{6}{s^2} + \frac{3}{s^2} + \frac{2}{s^2} + \frac{8}{s^3} + \frac{4}{s^3}\right) + \left(\frac{6}{s^3}\right)$$

$$\frac{x_1}{U} = \frac{2s + 5}{s^3 + 6s^2 + 19s + 26}$$

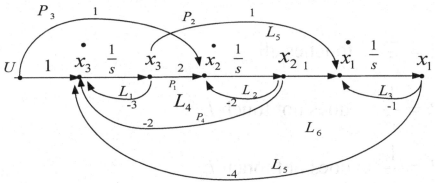

Figure 4.3 Example of a Relatively Complex Signal Flow Graph

$$L_5 = \frac{8}{s^3} \qquad \text{touches all}$$

$$L_6 = \frac{-4}{s^2} \qquad \text{does not touch } L_2$$

From the above loops, find the non-touching loops and form combinations taken two at a time.

The resulting five combinations are:

$$L_1 L_2 , \; L_1 L_3 , \; L_2 L_3 , \; L_2 L_6 , \; L_3 L_4 .$$

The non-touching loops taken three at a time is:

$$L_1 L_2 L_3 .$$

Next, find the forward paths. In particular the path, $P_4$, returned but went forward without traversing a loop. The gains of the paths are:

$$P_1 = \frac{2}{s^3} \qquad \text{touches all}$$

$$P_2 = \frac{1}{s^2} \qquad \text{does not touch } L_2$$

$$P_3 = \frac{1}{s^2} \qquad \text{does not touch } L_3$$

$$P_4 = -\frac{2}{s} \qquad \text{touches all}$$

The $\Delta$ from Mason's rule is found as follows:

$$\Delta=1+\left(\frac{6}{s}+\frac{8}{s^2}+\frac{8}{s^3}\right)+\left(\frac{6}{s^2}+\frac{3}{s^2}+\frac{2}{s^2}+\frac{8}{s^3}+\frac{4}{s^3}\right)+\left(\frac{6}{s^3}\right)$$

The numerator of the rule is:

$$\sum P_k\Delta_k=\frac{2}{s^3}+\frac{1}{s^2}\left(1+\frac{2}{s}\right)+\frac{1}{s^2}\left(1+\frac{3}{s}\right)-\frac{2}{s^3}$$

Finally, its transfer function is given by

$$\frac{x_1}{U}=\frac{2s+5}{s^3+6s^2+19s+26}.$$

In summary, Mason's rule is computationally intensive. The probability of committing an error during manual calculations is very high. It is, however, very rich in details. Readers interested in the approach may opt to develop a computer program in its applications.

# Chapter 5

# The Matrix Approach in Solving Differential Equations

The matrix approach is another way of solving control system problems. It applies to systems with or without initial conditions. In this sense, the matrix approach is a more general approach.

Transfer function does not apply to a system with initial conditions. Matrix approach is well suited to solve such a system.

## 5.1        BASICS OF THE MATRIX APPROACH

Consider a set of time-dependent linear differential equations

$$\dot{x}(t) = [A]x(t) + [B]u(t)$$

where,

$\dot{x}(t)$ = the differential equation as a function of time, $t$,
$[A]$ = coefficient matrix of $x(t)$, and
$[B]$ = coefficient matrix (vector) of input $u(t)$.

Taking the Laplace of both sides,

$$sx(s) - x(0) = [A]x(s) + [B]u(s)$$

where,

$x(0) =$ a matrix (vector) representing the initial conditions of $x$.

Collecting similar terms in $x(s)$ and factoring,

$$x(s)(s - [A]) = x(0) + [B]u(s).$$

Let $I$ be a unit matrix. An example of $I$ is

$$I = \begin{bmatrix} 1 & 0 & 0 \\ 0 & 1 & 0 \\ 0 & 0 & 1 \end{bmatrix}$$

Multiplying $s$ by the unit matrix gives

$$x(s)(s[I] - [A]) = x(0) + [B]u(s).$$

All factors and terms in the equation involving $x(s)$ are now matrices. It may be expressed as:

$$x(s) = (s[I] - [A])^{-1} x(0) + (s[I] - [A])^{-1} [B]u(s).$$

The inverse matrix, $(s[I] - [A])^{-1}$ is given by

$$(s[I] - [A])^{-1} = \frac{\text{adjoint } (s[I] - [A])}{\det (s[I] - [A])}.$$

Finding the adjoint of a matrix consists of two steps. First, it requires finding its transpose. Next, it requires finding the minor of its transpose. The flowing example illustrates the procedure.

Example 1. Find the transfer function, $\dfrac{x_1}{u}(s)$, of the following differential equations:

$$\dot{x}_1 = -7x_1 + x_2 + 2u$$

$$\dot{x}_2 = -14x_1 + x_3 + 6u$$

$$\dot{x}_3 = -8x_1 + 8u$$

The matrices [A] and [B] are

$$[A] = \begin{bmatrix} -7 & 1 & 0 \\ -14 & 0 & 1 \\ -8 & 0 & 0 \end{bmatrix}$$

and

$$[B] = \begin{bmatrix} 2 \\ 6 \\ 8 \end{bmatrix}.$$

Multiplying $s$ by the unit matrix,

$$s[I] = \begin{bmatrix} s & 0 & 0 \\ 0 & s & 0 \\ 0 & 0 & s \end{bmatrix}.$$

Subtracting [A] from the product gives

$$s[I]-[A]=\begin{bmatrix} s+7 & -1 & 0 \\ 14 & s & -1 \\ 8 & 0 & s \end{bmatrix}.$$

The determinant of $\left(s[I]-[A]\right)$ is

$$\det\left(s[I]-[A]\right)=s^3+7s^2+14s+8.$$

Its transpose is

$$\left(s[I]-[A]\right)^T=\begin{bmatrix} s+7 & 14 & 8 \\ -1 & s & 0 \\ 0 & -1 & s \end{bmatrix}.$$

Next, get the minors of the nine cells in the transposed matrix. First, replace a cell in the ith row and jth column by $(-1)^{i+j}$. Second, draw a vertical line and a horizontal line in the cell. The remaining cells are the elements of a new matrix. Do the scalar multiplication of $(-1)^{i+j}$ and the determinant of the resulting new matrix. The result is the minor of that ith and jth cell. The following illustrates the minors of the first four elements of the transpose above.

$$\text{minor of } s + 7 = (-1)^{1+1} \begin{bmatrix} s & 0 \\ -1 & s \end{bmatrix} = 1(s^2 - 0) = s^2$$

$$\text{minor of } 14 = (-1)^{1+2} \begin{bmatrix} -1 & 0 \\ 0 & s \end{bmatrix} = -1(-s) = s$$

$$\text{minor of } 8 = (-1)^{1+3} \begin{bmatrix} -1 & s \\ 0 & -1 \end{bmatrix} = 1(1-s) = 1-s$$

$$\text{minor of } -1 = (-1)^{2+1} \begin{bmatrix} 14 & 8 \\ -1 & s \end{bmatrix} = (-1)(14s - (-8)) = -14s - 8$$

and so on ...

Recall that the inverse matrix is

$$\left( s[I] - [A] \right)^{-1} = \frac{\text{adjoint} \left( s[I] - [A] \right)}{\det \left( s[I] - [A] \right)}.$$

Substituting the adjoint matrix and its determinant gives

$$\left( s[I] - [A] \right)^{-1} = \frac{\begin{bmatrix} s^2 & s & 1 \\ -14s - 8 & s^2 + 2s & s + 7 \\ -8s & 8 & s^2 + 7s + 14 \end{bmatrix}}{s^3 + 7s^2 + 14s + 8}.$$

Now, x(s) is

$$x(s) = \left( s[I] - [A] \right)^{-1} [B] u(s).$$

Substituting the [B] matrix,

$$x(s) = \left( s[I] - [A] \right)^{-1} \begin{bmatrix} 2 \\ 6 \\ 8 \end{bmatrix} u(s).$$

Performing the matrix multiplication gives the desired transfer function:

$$\frac{x_1}{u}(s) = \frac{2s^2 + 6s + 8}{s^3 + 7s^2 + 14s + 8}.$$

The other transfer function $\frac{x_2}{u}(s)$ or $\frac{x_3}{u}(s)$ may also be found in a similar way.

Example 2. Develop the set of differential equations representing the circuit shown on Figure 5.1.

Applying KVL and KCL,

$$V = R_1(i_1 + i_2) + V_C + R_2 i_1 + V_{L1}$$
$$V = R_1(i_1 + i_2) + V_C + R_3 i_2 + V_{L2}$$
$$i_C = i_1 + i_2.$$

Next, replace the voltage across an inductor by the value of inductance and the derivative of current through it. Do the same for current through a capacitor except use capacitance value and the derivative of voltage. The resulting set of equations follows:

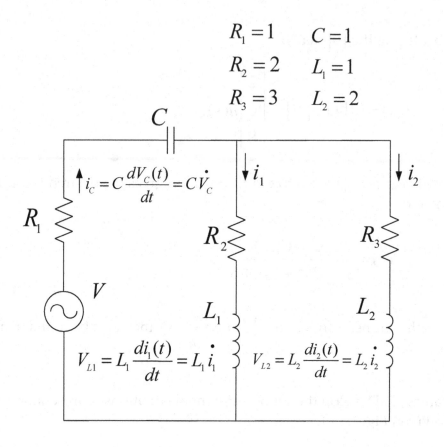

$$R_1 = 1 \qquad C = 1$$
$$R_2 = 2 \qquad L_1 = 1$$
$$R_3 = 3 \qquad L_2 = 2$$

USING KVL AND KCL:

$$V = R_1(i_1 + i_2) + V_C + R_2 i_1 + V_{L1}$$
$$V = R_1(i_1 + i_2) + V_C + R_3 i_2 + V_{L2}$$
$$i_C = i_1 + i_2$$

AFTER REPLACING VOLTAGE ACROSS AN INDUCTOR AND CURRENT THROUGH A CAPACITOR

$$V = R_1(i_1 + i_2) + V_C + R_2 i_1 + L_1 \dot{i}_1$$
$$V = R_1(i_1 + i_2) + V_C + R_3 i_2 + L_2 \dot{i}_2$$
$$C\dot{V}_C = i_1 + i_2.$$

Figure 5.1 Sample Circuit for Deriving the Set of Differential Equations

$$V = R_1(i_1 + i_2) + V_C + R_2 i_1 + L_1 \dot{i_1}$$

$$V = R_1(i_1 + i_2) + V_C + R_3 i_2 + L_2 \dot{i_2}$$

$$C\dot{V_C} = i_1 + i_2.$$

Substituting the following values for resistors, capacitor, and inductors:

$$R_1 = 1 \quad C = 1$$
$$R_2 = 2 \quad L_1 = 1$$
$$R_3 = 3 \quad L_2 = 2$$

After collecting terms and simplifying, the set of differentials equations (with state variables on the left hand side) is

$$\dot{i_1} = -3i_1 - i_2 - V_C + V$$

$$\dot{i_2} = -0.5i_1 - 2i_2 - 0.5V_C + V$$

$$\dot{V_C} = i_1 + i_2$$

## 5.2 REMARK ON THE EFFICIENCY OF MATRIX REPRESENTATION AND ITS LIMITATION

The efficiency of matrix representation lies on its capability to find state variables in one representation. Furthermore, it can be easily extended to systems with several inputs. For example, a system with two inputs can be represented by

$$x(s) = \left(s[I] - [A]\right)^{-1} x(0) + \left(s[I] - [A]\right)^{-1} [B_1] u_1(s) + \left(s[I] - [A]\right)^{-1} [B_2] u_2(s)$$

where $[B_1]$ and $[B_2]$ are coefficient matrix (vectors) of inputs $u_1$ and $u_2$, respectively.

In the design of electrical circuits, however, the interest is in the ratio of its output voltage and the input voltage. Oftentimes, the output element is a resistor. This means to get the voltage across the resistor, it must be multiplied by the current flowing it.

## 5.2.1    Remark on the application of the matrix methods on non-circuits

The matrix approach need not be applied on variables that are function of time. Independent variable may be anything. For example, in economic analysis, the independent variable maybe the savings rate of the population. The dependent variables may be the strength of the economy and standard of living.

# Chapter 6

# The Bode Plot

Bode plot graphically represents a transfer function's gain and phase (as a function of frequency). It is also useful in determining the gain and phase margins of a system. Bode plot can also be used in determining the stability of a system. Bode plot is perhaps is the most extensively used graphical system in representing a control system.

## 6.1     THE GAIN AND PHASE OF A COMPLEX NUMBER

Consider the complex number

$$c = a + jb.$$

Its magnitude or gain is

$$|c| = \sqrt{a^2 + b^2}$$

and its phase is the inverse tangent of the ratio of its imaginary part and real part,

$$\varphi = \tan^{-1}\left(\frac{b}{a}\right).$$

Note that two vertical bars represent the magnitude of a complex number. Because of the square operation, the formula for the magnitude applies independent of the sign of the real and imaginary parts. That is, if $c = a - jb$, then $|c| = \sqrt{a^2 + b^2}$ also. The same is true for $c = -a - jb$.

A complex number may also be represented in (polar) magnitude-phase, or phasor form:

$$c = a + jb = |c| \angle \varphi.$$

An important property of complex number that will be used in this chapter is the product of imaginary unit:

$$(j)(j) = j^2 = -1.$$

In electrical engineering, the same concept of magnitude and phase applies except that $c$ is the transfer function $G(s)$, $a$ is the cutoff frequency, and $b$ is simply the frequency of interest.

Obviously, if $G(s)$ is relatively complex, it pays to develop a relatively simple graphical representation of the transfer function. The representation must consist of the magnitude or gain plot, and the phase or angle plot. A Bode plot can provide such a representation.

## 6.2    ASYMPTOTIC REPRESENTATION OF A BODE PLOT

Examine the transfer function

$$G(s) = \frac{10000}{s(s+10)(s+100)}.$$

Let $s = j\omega$. Then $G(s)$ now becomes

$$G(\omega) = \frac{10000}{j\omega(j\omega+10)(j\omega+100)}.$$

The classical representation of Bode plot divides the frequency in a parenthesis by its cutoff frequency. This is not without a reason as will be seen later.

Performing the division,

$$G(\omega) = \frac{10000}{(10)(100)\,j\omega\left(\dfrac{j\omega}{10}+1\right)\left(\dfrac{j\omega}{100}+1\right)}.$$

Simplifying

$$G(\omega) = \frac{10}{j\omega\left(\dfrac{j\omega}{10}+1\right)\left(\dfrac{j\omega}{100}+1\right)}.$$

The last result has four factors as follows:

1.      10 (constant gain)

2.      $\dfrac{1}{j\omega}$ (real pole at the origin)

3.      $\dfrac{1}{\left(\dfrac{j\omega}{10}+1\right)}$ (real pole at $\omega = 10$)

4.      $\dfrac{1}{\left(\dfrac{j\omega}{100}+1\right)}$ (real pole at $\omega = 100$)

A zero is the opposite of a pole. It is a factor that appears in the numerator of a transfer function. If a real pole at the origin is $\dfrac{1}{j\omega}$, then the corresponding real zero at the origin is $j\omega$. Similarly, if the real pole at $\omega = 10$ is $\dfrac{1}{\left(\dfrac{j\omega}{10}+1\right)}$, its corresponding real zero at $\omega = 10$ is $\left(\dfrac{j\omega}{10}+1\right)$.

Bode plot uses the logarithm of the magnitude of $G(\omega)$. That is,

$$\log|G(w)| = \log 10 + \log\left|\frac{1}{j\omega}\right| + \log\left|\frac{1}{\left(\dfrac{j\omega}{10}+1\right)}\right| + \log\left|\frac{1}{\left(\dfrac{j\omega}{100}+1\right)}\right|.$$

In electrical engineering, a transfer function is the ratio of the output voltage and the input voltage. If the ratio is squared, its logarithm taken and multiplied by ten the result is

$$G(\omega) = 10\left(\log\left[\frac{V_{out}(\omega)}{V_{in}(\omega)}\right]^2\right).$$

Equivalently,

$$G(\omega) = 20\left(\log\frac{V_{out}(\omega)}{V_{in}(\omega)}\right) \text{ in dB}$$

To find the original ratio of voltages, simply take its inverse

$$\frac{V_{out}}{V_{in}} = 10^{\frac{G(\text{in dB})}{20}}.$$

Table 6.1 shows some typical values of the gain in dB given the ratio of the output voltage and the input voltage.

Recall the four factors above. It will be instructive to show how each factor can be represented in a Bode plot.

### 6.2.1   Magnitude of a constant gain (10 in the above example)

When the gain is constant, then it is true for all frequencies. That is the gain at $\omega = 0$, $\omega = 5$, or any other $\omega$'s should be the same. Tables 6.2(a) and table 6.2(b) show some examples of a constant gain. Figure 6.1 shows its Bode plot.

### 6.2.2   Magnitude of real zero or pole at the origin

Table 6.3(a) shows the variation of the gain with a real zero at the origin. For the real pole at the origin, the variation is shown on Table 6.3(b). Figure 6.2 shows its Bode plot.

### 6.2.3   Magnitude of real zero and real pole at $\omega = 10$

Table 6.4(a) and Table 6.4(b) show the variation of the gain with real zero and real pole, respectively. Figure 6.3 shows their Bode plots.

Note that the magnitude of a real zero and real pole at $\omega = 100$ will look the same except the cutoff frequency is at 100.

### 6.2.4   Composite Bode plot of all the factors

To get the composite plot of all the factors, simply add the values from each of the factor plot. Figure 6.4 shows the composite plot of all the factors.

Table 6.1 Variation in dB with some Ratios of output Voltage and Input Voltage

| $\dfrac{V_{out}}{V_{in}}$ | Expanded version of $\dfrac{V_{out}}{V_{in}}$ | $\log \dfrac{V_{out}}{V_{in}}$ | $20\left(\log \dfrac{V_{out}}{V_{in}}\right)$ dB |
|---|---|---|---|
| $10^5$ | 100000 | 5 | 100 |
| $10^4$ | 10000 | 4 | 80 |
| $10^3$ | 1000 | 3 | 60 |
| $10^2$ | 100 | 2 | 40 |
| $10^1$ | 10 | 1 | 20 |
| $10^0$ | 1 | 0 | 0 |
| $10^{-1}$ | 0.1 | -1.0 | -20 |
| $10^{-2}$ | 0.01 | -2.0 | -40 |
| $10^{-3}$ | 0.001 | -3.0 | -60 |
| $10^{-4}$ | 0.0001 | -4.0 | -80 |
| $10^{-5}$ | 0.00001 | -5.0 | -100 |
| $10^{-6}$ | 0.000001 | -6.0 | -120 |
| $10^{-7}$ | 0.0000001 | -7.0 | -140 |
| $10^{-8}$ | 0.00000001 | -8.0 | -160 |
| $10^{-9}$ | 0.000000001 | -9.0 | -180 |

Note: The logarithmic function is not defined for zero and negative numbers.

Table 6.2(a) Variation of dB with a Constant Gain (Gain is in the numerator)

| Constant gain $K$ | Expanded version of the constant $K$ | $20(\log K)$ dB |
|---|---|---|
| $10^3$ | 1000 | 60 |
| $10^2$ | 100 | 40 |
| $10^1$ | 10 | 20 |
| $10^0$ | 1 | 0 |

Table 6.2(b) Variation of dB with a Constant Gain (Gain is in the denominator)

| $K$ | Constant gain $\dfrac{1}{K}$ | $20(\log K)$ dB |
|---|---|---|
| $10^0$ | 1 | 0 |
| $10^1$ | 0.1 | -20 |
| $10^2$ | 0.01 | -40 |
| $10^3$ | 0.001 | -60 |

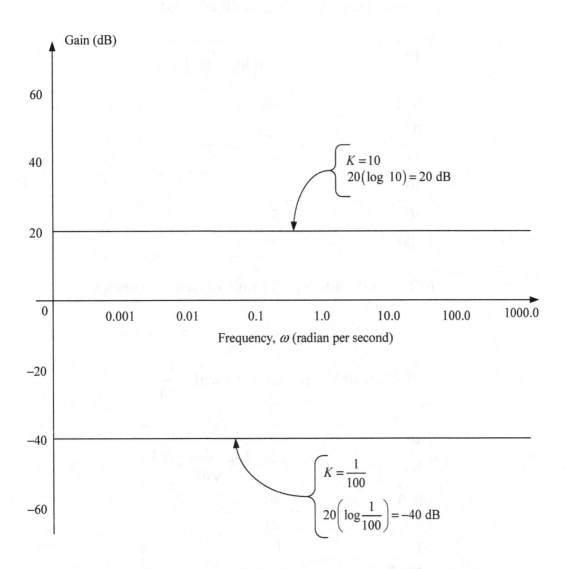

Figure 6.1 Bode Plot of some Constant Gains

Table 6.3(a) Variation in dB with $j\omega$

| $\omega$ | $20\left(\log\sqrt{\omega^2}\right)$ dB |
|---|---|
| 0.01 | -40 |
| 0.1 | -20 |
| 1 | 0 |
| 10 | 20 |
| 100 | 40 |
| 1000 | 60 |

NOTE: A decade is a 10:1 ratio of two frequencies.

Table 6.3(b) Variation in dB with $\dfrac{1}{j\omega}$

| $\omega$ | $20\left(\log\dfrac{1}{\sqrt{\omega^2}}\right)$ dB |
|---|---|
| 0.01 | 40 |
| 0.1 | 20 |
| 1 | 0 |
| 10 | -20 |
| 100 | -40 |
| 1000 | -60 |

NOTE: A decade is a 10:1 ratio of two frequencies.

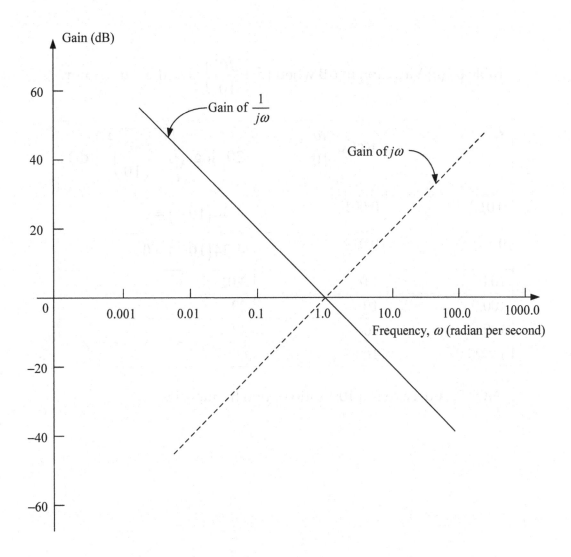

Figure 6.2 Bode Plot of a Real Zero and Real Pole at the Origin

Table 6.4(a) Variation in dB when $\left(1 + \dfrac{j\omega}{10}\right)$ is in the Numerator

| $\omega$ | Ratio $\dfrac{\omega}{10}$ | $20\left(\log \sqrt{1 + \left(\dfrac{\omega}{10}\right)^2}\right)$ dB |
|---|---|---|
| 0.01 | 0.001 | $4.34\left(10^{-6}\right) \approx 0$ |
| 0.1 | 0.01 | $4.34\left(10^{-4}\right) \approx 0$ |
| 10.0 | 1.0 | 3.01 |
| 100.0 | 10 | 20 |
| 1000.0 | 100 | 40 |
| 10000.0 | 1000 | 60 |

NOTE: A decade is a 10:1 ratio of two frequencies.

Table 6.4(b) Variation in dB when $\left(1+\dfrac{j\omega}{10}\right)$ is in the Denominator

| $\omega$ | Ratio $\dfrac{\omega}{10}$ | $20\left|\log \dfrac{1}{\sqrt{1+\left(\dfrac{\omega}{10}\right)^2}}\right|$ dB |
|---|---|---|
| 0.01 | 0.001 | $4.34\left(10^{-6}\right)\approx 0$ |
| 0.1 | 0.01 | $4.34\left(10^{-4}\right)\approx 0$ |
| 10.0 | 1 | -3.01 |
| 100.0 | 10 | -20 |
| 1000.0 | 100 | -40 |
| 10000.0 | 1000 | -60 |

NOTE: A decade is a 10:1 ratio of two frequencies.

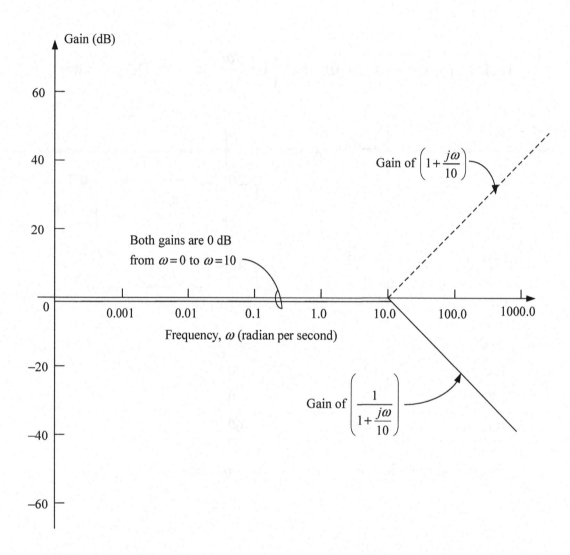

Figure 6.3 Bode Plot of a Real Zero and Real Pole at Cutoff Frequency

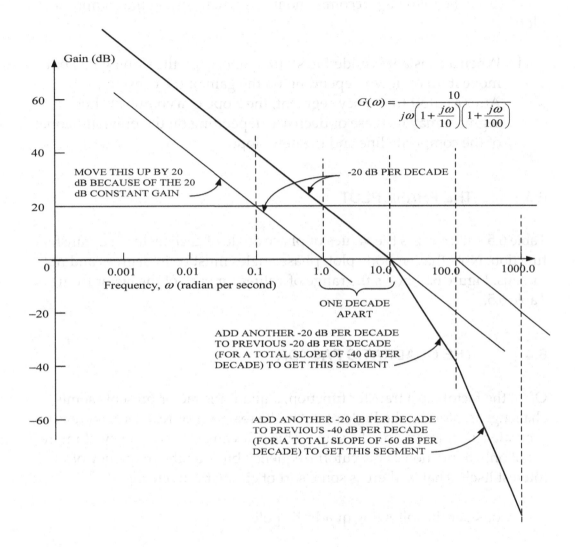

Figure 6.4  The Composite or Resulting Plot of all the Factor Plots

Observe the following recommendations when getting the composite plot:

1. When a constant is added to sloping line, copy the sloping line but move it up or down depending on the gain of the constant,
2. After a cutoff frequency segment, the slope of a composite line segment may increase or decrease depending on the previous slope of the composite line and the new slope.

## 6.3    THE PHASE PLOT

Table 6.5 summarizes the values of phase angle of each factor in a transfer function. Note that, in Bode plot, phase angles must be in degrees and not radians. Figure 6.5 shows the range of values for each of the factor from Table 6.5.

## 6.4    THE QUADRATIC FACTOR

Of all the factors in a transfer function, a quadratic factor presents some challenge in plotting the Bode plot. Unlike real pole or real zero, there is no accurate rule for drawing the slope. This is because the frequency of interest is not only a function of the cutoff frequency but also the frequency of interest itself. That is, there is some sort of circular referencing.

Consider the following quadratic pole:

$$G(s) = \frac{1}{s^2 + bs + c}.$$

The interest is to find the parameters $s$, $b$, and $c$ such that

$|G(s)|=1$.

First, replace $G(s)$ by $G(\omega)$:

Table 6.5 Variation in the Phase Angle Depending on the Type of Factor and its Location in a Transfer Function

| Type of factor | Phase angle in degrees |
|---|---|
| Positive constant such as 10 | 0 |
| Negative constant such as - 10 | $\pm180$ |
| $j\omega$ | 90 |
| $\dfrac{1}{j\omega}$ | -90 |
| $\left(1+\dfrac{j\omega}{10}\right)$ | From zero to 90 degrees with 45 degrees at $\omega=10$. |
| $\dfrac{1}{\left(1+\dfrac{j\omega}{10}\right)}$ | From zero to -90 degrees with -45 degrees at $\omega=10$. |
| $\left(1+\dfrac{j\omega}{100}\right)$ | From zero to 90 degrees with 45 degrees at $\omega=100$. |
| $\dfrac{1}{\left(1+\dfrac{j\omega}{100}\right)}$ | From zero to -90 degrees with -45 degrees at $\omega=100$. |

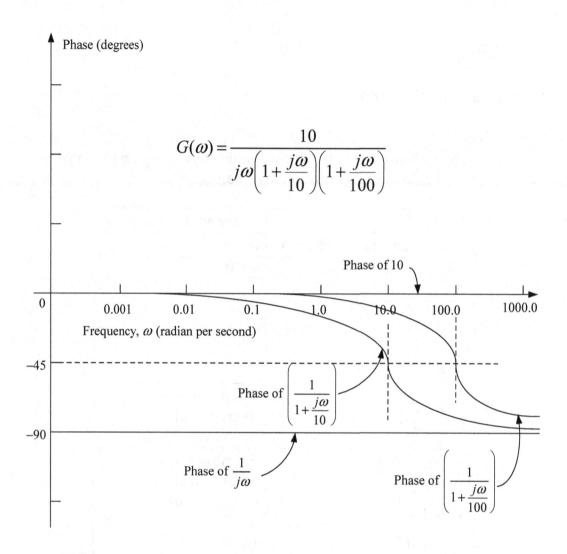

Figure 6.5 Phases of Different Factors in the Transfer Function

$$G(\omega) = \frac{1}{(j\omega)^2 + b(j\omega) + c}$$

The natural or cutoff frequency is $\omega = \sqrt{c}$. Substituting the same to the equation,

$$G(\omega) = \frac{1}{-\left(\sqrt{c}\right)^2 + jb\sqrt{c} + c}$$

or,

$$G(\omega) = \frac{1}{jb\sqrt{c}}.$$

Its magnitude is

$$|G(\omega)| = \left|\frac{1}{jb\sqrt{c}}\right|.$$

To satisfy $|G(s)| = 1$,

$$b = \left|\frac{1}{\sqrt{c}}\right|$$

or

$$b = \frac{\sqrt{c}}{c}.$$

Note: To satisfy a gain of $|G(\omega)| = K$, $b = \frac{K\sqrt{c}}{c}$.

Hence, $G(\omega) = \dfrac{1}{jb\sqrt{c}}$ becomes $|G(\omega)| = \dfrac{1}{j1} = 1$. Now, using $b = \dfrac{\sqrt{c}}{c}$ in the original equation:

$$G(\omega) = \frac{1}{(j\omega)^2 + b(j\omega) + c}$$

gives

$$G(\omega) = \frac{1}{\dfrac{\sqrt{c}}{c}(j\omega) + c - \omega^2}.$$

Examining the imaginary and real parts of the denominator of the last equation shows that $\omega$ from $j\omega$ depends on $c - \omega^2$. That is, the frequency of interest depends on the cutoff or natural frequency and the frequency itself. If the frequency of interest is one decade away, or ten times the cutoff or natural frequency

$$|G(\omega)| = \frac{1}{\sqrt{10^2 + (99c)^2}}.$$

This is in contrast with a simple pole such as $\dfrac{1}{\left(\dfrac{j\omega}{10} + 1\right)}$. In this case, if the frequency of interest is ten times the cutoff frequency then its gain

is $\dfrac{1}{\left(\dfrac{j100}{10} + 1\right)} = \dfrac{1}{j10 + 1} = 0.0995$ (or -20 dB), which is constant.

The best approach in determining the shape of the quadratic factor is to substitute various values of frequency and solve for the gain and phase. Table 6.6(a) shows the sample calculations for quadratic zero and Table 6.6(b) for the quadratic pole. Figure 6.7 shows the Bode plots of each table.

Table 6.6(a) Sample Calculations for the Gain of a Quadratic Zero

Given: $G(s) = s^2 + \dfrac{10}{100}s + 100$ then $G(\omega) = -\omega^2 + j0.1\omega + 100$

| $\omega$ | $20\left(\log|G(\omega)|\right)$ dB |
|----------|-------------------------------------|
| 0        | 40.0                                |
| 1        | 40.1                                |
| 6        | 36.12                               |
| 10       | 0.0                                 |
| 20       | 49.5                                |
| 80       | 76.0                                |
| 100      | 79.9                                |

## 6.4.1    Role of quadratic factor in time domain

As will be seen in the next chapter, the quadratic factor is responsible for some initial oscillatory response in its time domain. That is, the quadratic factor explains events such as overshoot, rise time, and settling time. In a general case, a quadratic factor is the dominant pole and controls the stability of a system.

Table 6.6(b) Sample Calculations for the Gain of a Quadratic Pole

Given: $G(s) = \dfrac{1}{s^2 + \dfrac{10}{100}s + 100}$ then $G(\omega) = \dfrac{1}{-\omega^2 + j0.1\omega + 100}$

| $\omega$ | $20\big(\log|G(\omega)|\big)$ dB |
|---|---|
| 0 | -40.0 |
| 1 | -40.1 |
| 6 | -36.12 |
| 10 | 0.0 |
| 20 | -49.5 |
| 80 | -76.0 |
| 100 | -79.9 |

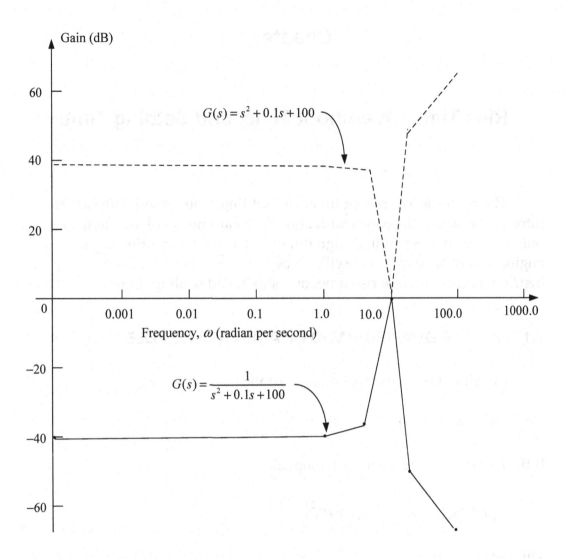

Figure 6.6 Sample Plot for a Quadratic Zero and Quadratic Pole

# Chapter 7

# Rise Time, Overshoot Time, and Settling Time

The rise time, overshoot time, and settling time provide important information about the bandwidth and other parameters of a system. Not only are they important in design but in test as well. Oftentimes, an engineer may have only an oscilloscope to test a system. Such an instrument can measure rise time, overshoot, and settling times.

## 7.1      THE DAMPING RATIO OF A QUADRATIC POLE

Consider taking the root of a quadratic factor. That is,

$$s^2 + bs + c = 0.$$

If the factor is a perfect square trinomial,

$$(s + \omega_0)^2 = s^2 + 2\omega_0 s + \omega_0^2.$$

This means

$$b = 2\omega_0 \text{ and } c = \omega_0^2.$$

Generalize the quadratic factor by multiplying $b$ by $\zeta$. That is,

$$s^2 + 2\zeta\omega_0 s + \omega_0^2 = 0$$

with

$$b = 2\zeta\omega_0.$$

$\zeta$ is called damping ratio and equal to:

$$\zeta = \frac{b}{2\omega_0}.$$

By using the quadratic formula, the two roots of $s$ are:

$$s = \omega_0\left(-\zeta \pm \sqrt{\zeta^2 - 1}\right).$$

When $\zeta = 1$,

$$s = -\omega_0\zeta.$$

The corresponding factored form of $s^2 + 2\zeta\omega_0 s + \omega_0^2$ is:

$$s^2 + 2\zeta\omega_0 s + \omega_0^2 = (s + \omega_0\zeta)(s + \omega_0\zeta).$$

When $\zeta > 1$

$$s = \omega_0\left(-\zeta \pm a\right)$$

which, correspond to the factored form

$$s^2 + 2\zeta\omega_0 s + \omega_0^2 = (s + \omega_0\zeta + \omega_0 a)(s + \omega_0\zeta - \omega_0 a).$$

Finally, when $\zeta < 1$,

$$s = \omega_0 \left( -\zeta \pm jb \right)$$

corresponding to

$$s^2 + 2\zeta\omega_0 s + \omega_0^2 = \left( s + \omega_0\zeta + j\omega_0 b \right)\left( s + \omega_0\zeta - j\omega_0 b \right).$$

The first case, $\zeta = 1$, is critical damping. A critically damped system settles the fastest towards the settling time. It has no overshoot. Critical damping is oftentimes desirable.

$\zeta > 1$ corresponds to the overdamped case. Like critical damping, it has two real poles. The poles, however, are not equal. It takes a long time for an overdamped case to settle.

When $\zeta < 1$, the roots are complex (with real and imaginary parts). It is the underdamped case. Underdamped behavior has oscillatory response and overshoot. Its settling time is between the critically damped and overdamped cases. Its rise time, however, is the smallest among the three.

Figures 7.1(a), (b), and (c) show a sample transfer function of each case together with two other real poles. In addition, the figure shows the location of the poles on the complex plane. Figure 7.2 shows the variation in time of each case.

## 7.2    RISE TIME

Various definitions of rise time exist. For example, rise time in digital system is defined from zero to 50% of the steady state value. In control system, it is from zero to 70.7% of the steady state value. The 70.7% voltage gain corresponds to a power gain of 0.50 (or -3.01 dB). Figure 7.3 shows the locations of rise time together with overshoot time and settling time.

Independent of its definition, rise time is inversely proportional to the bandwidth. Larger bandwidth equate to smaller or faster rise time.

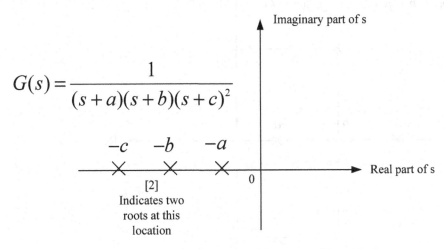

$$G(s) = \frac{1}{(s+a)(s+b)(s+c)^2}$$

(b) Location of the Roots of Case I in the Complex Plane

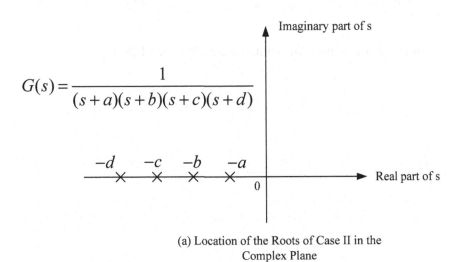

$$G(s) = \frac{1}{(s+a)(s+b)(s+c)(s+d)}$$

(a) Location of the Roots of Case II in the Complex Plane

Figure 7.1 Locations of Roots in the Complex Plane

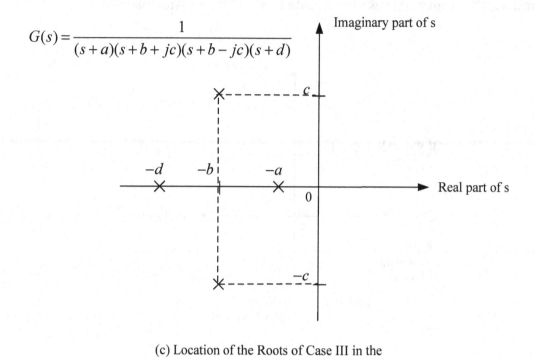

$$G(s) = \frac{1}{(s+a)(s+b+jc)(s+b-jc)(s+d)}$$

(c) Location of the Roots of Case III in the
Complex Plane

Figure 7.1(continued) Location of Roots in the Complex Plane

NOTES:

1.  CASE I CORRESPONDS TO CRITICAL DAMPING.  IT SETTLES THE EARLIEST.  CASE I HAS DOUBLE POLES.

2.  CASE II IS THE OVERDAMPED CASE.  THE RESPONSE TAKES THE LONGEST TO SETTLE.

3.  CASE III IS THE UNDERDAMPED CASE.  ITS COMPLEX ROOTS ARE RESPONSIBLE FOR THE OVERSHOOT AND OSCILLATORY RESPONSE.  THIS CASE HAS ALSO THE SMALLEST RISE TIME INDICATING A LARGER BANDWIDTH.

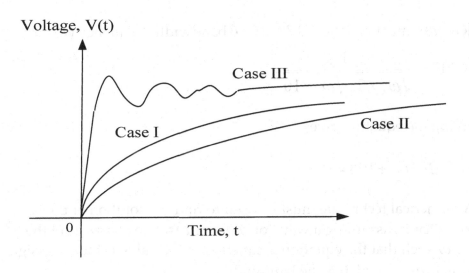

Figure 7.2 Time Response of Cases I, II, and III

Perhaps, one of the best applications of the Bode plot is in estimating the -3.01 dB bandwidth. While it can be calculated, the calculations require numerical techniques. Consider for example a system whose transfer function is:

$$G(s) = \frac{1}{(s+2)(s+10)}.$$

Getting its magnitude,

$$|G(j\omega)| = \sqrt{\frac{1}{\omega^2 + 10^2}} = \frac{1}{\sqrt{\omega^2 + 2^2}\sqrt{\omega^2 + 10^2}}.$$

Since $|G(j\omega)|$ must equal to 70.7% at the bandwidth frequency ($\omega = \omega_{-3dB}$)

$$0.707 = \frac{1}{\sqrt{\omega^2 + 2^2}\sqrt{\omega^2 + 10^2}}.$$

Simplifying the equation gives

$$\omega^4 + 204\omega^2 + 198 = 0.$$

A numerical technique must be used to find the solution of the last equation. That is, assuming a value of $\omega$ from $\omega = 0$ to $\omega = \infty$, find the value of $\omega$ such that the equation is satisfied. If the value is found, assign that value to $\omega_{-3dB}$, which is the bandwidth.

Of course, the number of calculations may be minimized by assuming a starting value of $\omega$ close to the smallest pole (e.g. $\omega = 2$) and ending at largest pole (e.g. $\omega = 10$). In any case, knowledge of its Bode plot will be extremely helpful.

## 7.3     OVERSHOOT TIME

Overshoot is caused by the imaginary part of the dominant complex pole, and the sum of the angles from the real poles and the real zeros. It is given by

$$t_p = \frac{1}{\omega_d} \left[ \frac{90^0 + 90^0 + \beta - \gamma}{57.3^0 \text{ / radian}} \right]$$

where

$\omega_d$ = imaginary part of the dominant complex pole,

$\beta$ = sum of all the angles from the real poles to the dominant complex pole, and

$\gamma$ = sum of all the angles from real zeros to the dominant complex pole.

To find the angle from a real zero or a real pole, connect a line from the zero or pole to the dominant complex pole and find its slope. Next, get its inverse tangent.

## 7.4     SETTLING TIME

The settling time depends on the real part of the dominant complex pole and the accuracy of a meter used in capturing its step response. Assume for example a transfer function of the form

$$\frac{C}{R}(s) = \frac{80}{(s+10)(s^2 + 4s + 8)}.$$

Let $R$ be a unit step with Laplace transform $R(s) = \frac{1}{s}$ (see chapter 1).

Factoring the quadratic pole,

$$s^2 + 4s + 8 = (s + 2 - j2)(s + 2 + j2).$$

$C(s)$ may now be expressed as

$$C(s) = \frac{80}{s(s+10)(s+2 \pm j2)}$$

where $R(s)$ was used in multiplying both sides of the original transfer function.

Using partial fraction expansion,

$$C(s) = \frac{A}{s} + \frac{B}{s+10} + \frac{D}{s+2-j2} + \frac{D^*}{s+2+j2}.$$

Next, solve for the constants in the numerator of each fraction:

$$A = sC(s)\big|_{s=0} = 1$$

$$B = (s+10)C(s)\big|_{s=-10} = 0.11$$

$$D = (s+2-j2)C(s)\big|_{s=-2+j2=\delta_d+\omega_d} = D\angle\phi_d = 0.9\angle121^0.$$

The partial fraction form of $C(s)$ is now given by

$$C(s) = \frac{1}{s} + \frac{0.11}{s+10} + \frac{0.9\angle121^0}{s+2-j2} + \frac{D^*}{s+2+j2}.$$

Using the inverse Laplace transform (converts from s domain to time domain),

$$C(t) = 1 - 0.11e^{-10t} + C_d(t)$$

where

$$C_d(t) = 2|D|e^{\sigma_d t}\cos(\omega_d t + \phi_d)$$

$$= 2|0.9|e^{-2t}\cos(2t + 121^0).$$

$C_d(t)$ is the time-domain response of the dominant complex pole.

Note that $\sigma_d$ is the real part of the dominant complex poles $(s+2-j2)(s+2+j2)$.

Table 7.1 shows the exponential decay of its product with time, $t$. As the product of $\sigma_d t$ becomes more negative, its exponential function decreases.

Consider a meter with 3% accuracy. Then the evaluation of the settling time requires an accuracy of better than 3%. In this case, use 1.83% (from Table 1).

Table 7.1 Variation of the Exponential Function of $\sigma_d t$

| $\delta_d t$ (no unit) | $e^{\delta_d t}$ (no unit) | Percent (%) |
| --- | --- | --- |
| 0 | 1 | 100.0 |
| -1 | 0.3678 | 36.78 |
| -2 | 0.1353 | 13.53 |
| -3 | 0.0498 | 4.98 |
| -4 | 0.0183 | 1.83 |

Next, derive the settling time as follows:

$$e^{\delta_d t} = e^{\delta_d t}\Big|_{t=t_s} = e^{\delta_d t_s} = e^{-m}$$

where

$m$ = accuracy of the meter in decimal.

Continuing the derivation,

$$\delta_d t_s = -m$$

or,

$$t_s = \frac{-m}{\delta_d}.$$

For the example above,

$$e^{-2t_s} = e^{-4}$$

or

$$t_s = \frac{4}{2} = 2 \text{ seconds.}$$

# Chapter 8

# The Realization of Analog Control Systems

Given a transfer function, its signal flow graph and realization diagram can be found. A realization diagram consists of four parts only. They are potentiometers, inverting summers, integrators, and inverters. Any analog control system may be realized using such parts.

## 8.1    BASIC COMPONENTS OR PARTS OF A REALIZATION DIAGRAM

The main building blocks of an analog electronic circuit design consist of potentiometer, inverter, inverting summer, and inverting integrator. Figure 8.1 and Figure 8.2 show the schematic diagrams of the blocks. A potentiometer is simply a voltage divider.

The input of an operational amplifier has a virtual ground. This is the key in deriving the voltage gains of an inverter, inverting summer, and inverting integrator. By applying Kirchoff's current law on the virtual ground, the gain of any of the above device can be found. In all cases, the magnitude of the voltage gain is the ratio of the feedback resistor or capacitor and the resistor in series with the input voltage.

In synthesis or realization diagrams, each of the above blocks are represented in a more compact form. For example, a potentiometer is a circle with the ratio of the voltages next to it. Similarly, an inverter is simply shown as a triangle, an inverting summer as a triangle with several inputs, and a combination of a rectangle and triangle for the inverting integrator. Figure 8.3 shows such a representation for a potentiometer, inverter, and

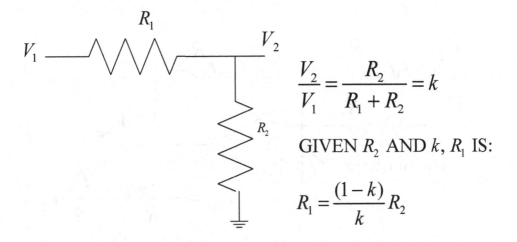

$$\frac{V_2}{V_1} = \frac{R_2}{R_1 + R_2} = k$$

GIVEN $R_2$ AND $k$, $R_1$ IS:

$$R_1 = \frac{(1-k)}{k} R_2$$

(a) Potentiometer as a voltage divider

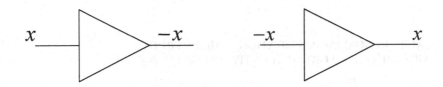

(b) Inverters with two types of inputs

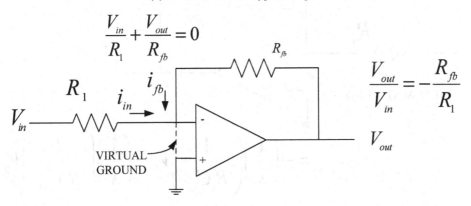

$$\frac{V_{in}}{R_1} + \frac{V_{out}}{R_{fb}} = 0$$

$$\frac{V_{out}}{V_{in}} = -\frac{R_{fb}}{R_1}$$

(c) Schematic diagram of an inverter

Figure 8.1 Schematic Diagrams of a Potentiometer and an Inverter

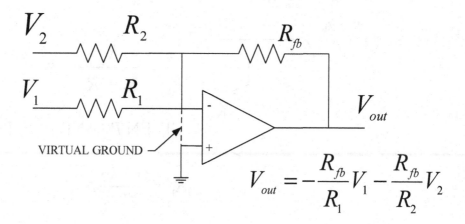

(a) Schematic diagram of an inverting summer

NOTE:  IN NORMAL APPLICATIONS, THE RC PRODUCT
MUST BE ONE TO OBTAIN A UNTIY INTEGRATION OF -1/s.

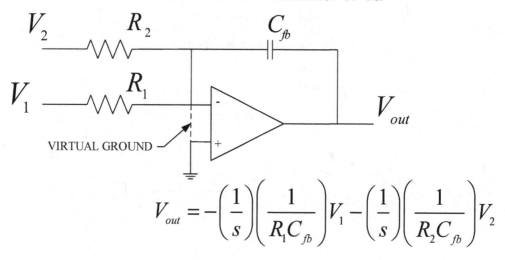

(b) Schematic diagram of an inverting integrator

Figure 8.2 Schematic Diagrams of an Inverting Summer and Inverting Integrator

THE "10" REPRESENTS THE VOLTAGE GAIN IN
THE INVERTING SUMMER.

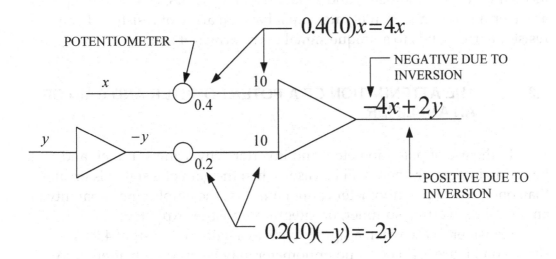

Figure 8.3 Compact Representations of Potentiometer, Inverter, and
Inverting Summer

inverting summer. An inverting integrator may be used in a similar manner
as an inverting summer.

## 8.2     SIGNALS IN THE PICKOFF POINT VERSUS POINT OF CONVERGENCE

A pickoff point is a point with one and exactly one signal. It is usually
the output of a device. Feedback signals starts at a pickoff point and ends at
the input of another device.

The point of convergence is where two or more signals converge. On a signal flow graph, such a point is shown as a node. This can be confusing since no two signals must terminate in a node.

When signals converge on the input of a device, each signal must have its own wire, potentiometer, and separate gain resistors. Hence, the input of an inverting summer or integrator must have an array of resistors. Each resistor corresponds to a unique signal in the array (of signals).

## 8.3 THE ATTENUATION OF A POTENTIOMETER AND GAIN OF AN AMPLIFIER

In the use of potentiometers and inverters, inverting summer, and inverting integrator, there will be cases when the gain of a signal is greater than one. Since a potentiometer cannot amplify, the amplifying capability of an inverter, inverting summer, or integrator must be exploited.

Consider a case when a signal requires a gain (or boost) of 4.0 as shown on Figure 8.4. Then a potentiometer may be used with attenuation ratio of 0.4. To compensate for the attenuation, an amplifier must have a gain of 10.0. That is, the required gain of $4.0 = (0.4)(10.0)$.

## 8.4 MATHEMATICS OF AN INVERTING SUMMER AS IT APPLIES TO SIGNAL FLOW GRAPH

In developing the realization of a signal flow graph, it is best to think the sign of the feedback signal to be the sign of all integrators and constants in the forward path. If the sign of the forward path is negative and the gain of the feedback path is positive then it implies that the sign of the feedback is negative. For example, in the loop with a gain of 8, the sign of the feedback is negative because the sign of its forward path is negative.

An inverting summer automatically takes into consideration the negative nature of the feedback signal. At the input of the inverting summer is a negative signal. Its output is positive. This (positive) result is then multiplied by the product of the integrators and constants in the

$$\frac{C}{R}(s) = \frac{4s + 32}{s^3 + 3s^2 + 22s + 48}$$

$$\frac{C}{R}(s) = \frac{\dfrac{4}{s^2} + \dfrac{32}{s^3}}{1 - \left(-\dfrac{3}{s} - \dfrac{22}{s^2} - \dfrac{48}{s^3}\right)}$$

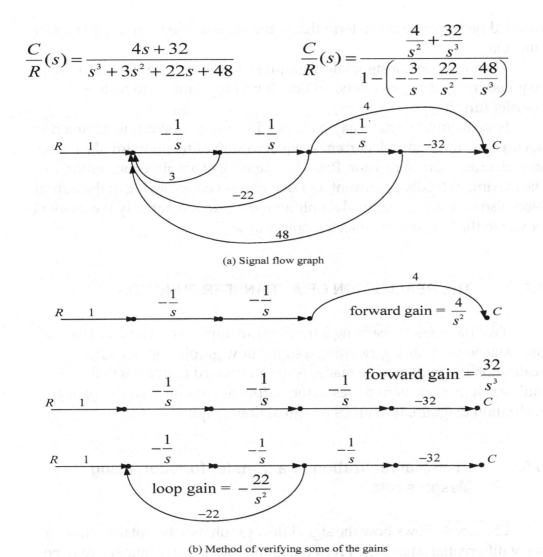

(a) Signal flow graph

(b) Method of verifying some of the gains

Figure 8.4 Signal Flow Graph using Mason's Rule

forward path resulting in a term that is the same as the term in the transfer function.

Therefore, once a signal flow graph is developed, simply insert the required integrators, inverters, and the inverting summer to realize the transfer function.

Several input signals may be applied to a single inverting summer. In such a case, the feedback current is equal to sum of the individual currents in each input (series) resistor. Rate the operational amplifier according to the maximum feedback current and maximum voltage swing at the output. Note also that the operational amplifier must provide not only the incident power to the load but its reflected power as well.

## 8.5      THE REALIZATION OF A TRANSFER FUNCTION

Two methods of realizing a transfer function is shown here. The first uses Mason's rule and generating a signal flow graph from which a realization diagram may be made. A recommended method is that of the author's. It uses the derivatives of the output as a guide in developing a realization diagram. It requires no signal flow graph.

## 8.5.1      Classical realization of a transfer function using Mason's rule

Chapter 4 shows how the signal flow graph may be obtained from a set of differential equations. From the graph, the transfer function may be obtained using Mason's rule. In Mason's rule, the numerator of the transfer function consists of terms in the forward path. Conversely, the denominator of the transfer function consists of terms in the feedback path.

In synthesis, the transfer function is known. Mason's rule is then applied to reconstruct the signal flow graph of the transfer function. Once the signal flow graph is known, components, such as potentiometers and inverter, may be used to realize the synthesis.

As an example, consider synthesizing the following transfer function:

$$\frac{C}{R}(s) = \frac{4s + 32}{s^3 + 3s^2 + 22s + 48}.$$

To format the transfer function according to Mason's rule, first divide the numerator and the denominator by the highest power of $s$. This gives,

$$\frac{C}{R}(s) = \frac{\dfrac{4s}{s^3} + \dfrac{32}{s^3}}{\dfrac{s^3}{s^3} + \dfrac{3s^2}{s^3} + \dfrac{22s}{s^3} + \dfrac{48}{s^3}}$$

$$= \frac{\dfrac{4}{s^2} + \dfrac{32}{s^3}}{1 + \dfrac{3}{s} + \dfrac{22}{s^2} + \dfrac{48}{s^3}}.$$

Finally, format the denominator such that it appears as in Mason's rule. The result follows:

$$\frac{C}{R}(s) = \frac{\dfrac{4}{s^2} + \dfrac{32}{s^3}}{1 - \left( -\dfrac{3}{s} - \dfrac{22}{s^2} - \dfrac{48}{s^3} \right)}.$$

The signal flow graph of the transfer function is shown on Figure 8.5. The figure also shows how some of the forward and loop gains may be verified.

Figure 8.6 shows the realization of the synthesis. It consists of three integrators, four potentiometers, one inverter, and one inverting summer. Notice how potentiometers are scaled down between 0.1 and 1.0, and how

$$\frac{C}{R}(s) = \frac{4s+32}{s^3+3s^2+22s+48} \qquad \frac{C}{R}(s) = \frac{\dfrac{4}{s^2}+\dfrac{32}{s^3}}{1-\left(-\dfrac{3}{s}-\dfrac{22}{s^2}-\dfrac{48}{s^3}\right)}$$

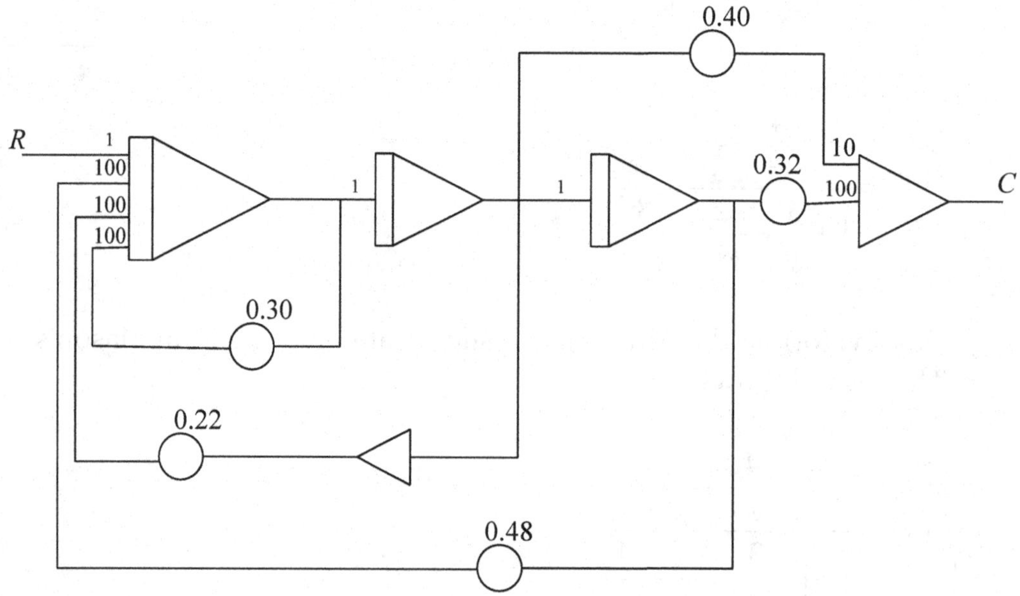

Figure 8.5 Realization of the Signal Flow Graph shown on Figure 8.4

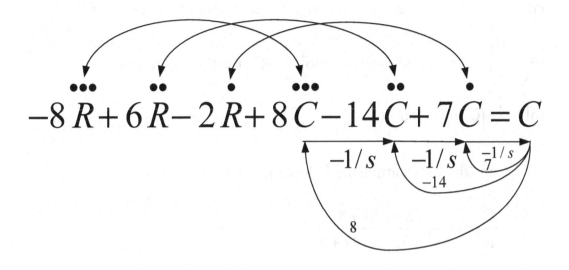

Figure 8.6 Representation of the Signal Flow Graph on the $C$ Equation

gains of operational amplifiers are scaled up by 10 or 100. Two of the integrators have gains of 1.0. As much as possible use a gain of 1.0 for integrators since most transfer functions use such a gain.

## 8.5.2    Realization using the derivatives of $C(s)$

In this approach, the numerator and denominator of the transfer function are cross multiplied first. Next, the output variable is solved by dividing both sides of the equation by the factor of s with the highest exponent. Inspection of each term follows. If a term has odd power of $s$, a negative sign is placed before $\dfrac{1}{s}$. Conversely, a positive sign is used for even power of $s$. If a change occurs, the sign of the coefficient must also change to its opposite. Finally, assign the first derivative of $C$ (or $\dot{C}$)

to $\left(-\dfrac{1}{s}\right)C$, its second derivative to $\left(\dfrac{1}{s^2}\right)C$, the third derivative

to $\left(-\dfrac{1}{s^3}\right)C$ and so on.

Consider for example the following transfer function:

$$\frac{C}{R}(s) = \frac{2s^2 + 6s + 8}{s^3 + 7s^2 + 14s + 8}.$$

Cross-multiplying gives

$$s^3 C + 7s^2 C + 14sC + 8C = 2s^2 R + 6sR + 8R.$$

Dividing both sides of the equation by the highest power of s,

$$\frac{s^3}{s^3}C + 7\left(\frac{s^2}{s^3}\right)C + 14\left(\frac{s}{s^3}\right)C + 8\left(\frac{1}{s^3}\right)C = 2\left(\frac{s^2}{s^3}\right)R + 6\left(\frac{s}{s^3}\right)R + 8\left(\frac{1}{s^3}\right)R$$

or,

$$C = -7\left(\frac{1}{s}\right)C - 14\left(\frac{1}{s^2}\right)C - 8\left(\frac{1}{s^3}\right)C + 2\left(\frac{1}{s}\right)R + 6\left(\frac{1}{s^2}\right)R + 8\left(\frac{1}{s^3}\right)R$$

Next, change the sign of each term such that odd power of s has negative sign and even power of s has positive sign. Change the sign of the coefficient to maintain the equality. Do the same for $R$.

$$C = 7\left(-\frac{1}{s}\right)C - 14\left(\frac{1}{s^2}\right)C + 8\left(-\frac{1}{s^3}\right)C - 2\left(-\frac{1}{s}\right)R + 6\left(\frac{1}{s^2}\right)R - 8\left(-\frac{1}{s^3}\right)R$$

The assignment of derivatives follows:

$$C = 7\dot{C} - 14\ddot{C} + 8\dddot{C} - 2\dot{R} + 6\ddot{R} - 8\dddot{R}.$$

Re-arranging the terms for the output to appear on the right hand side,

$$-8\dddot{R} + 6\ddot{R} - 2\dot{R} + 8\dddot{C} - 14\ddot{C} + 7\dot{C} = C.$$

All terms with variable C are feedback terms. Terms with R are forward terms. If a term has a negative coefficient then that coefficient must be represented by an inverter. A path involving a derivative must be represented by $-\frac{1}{s}$.

Figure 8.7 shows the equation for $C$ together with the signal flow graph superimposed on the equation. It is not the signal flow graph yet.

To draw the signal flow graph, notice that $\dot{R}$ must be equal to $\dot{C}$, $\ddot{R}$ to $\ddot{C}$, and $\dddot{R}$ to $\dddot{C}$. That is, for the input $R$ to obtain $\dot{R}$, it must be operated on by $\dot{C}$, and so on. Hence, all derivatives of $R$, e.g. $\dot{R}$, $\ddot{R}$, and $\dddot{R}$, may be replaced by $R$ provided $R$ is connected to the corresponding derivative of $C$.

Figure 8.8 is the resulting signal flow graph. Note that feedback signals are diverging from the output. This is called type I synthesis or realization.

A method of verifying each forward gain is shown on Figure 8.9. Each loop gain may be verified as shown on Figure 8.10.

$$\frac{C}{R}(s) = \frac{2s^2 + 6s + 8}{s^3 + 7s^2 + 14s + 8}$$

$$C = 7\left(-\frac{1}{s}\right)C - 14\left(\frac{1}{s^2}\right)C + 8\left(-\frac{1}{s^3}\right)C - 2\left(-\frac{1}{s}\right)R + 6\left(\frac{1}{s^2}\right)R - 8\left(-\frac{1}{s^3}\right)R$$

$$C = 7\dot{C} - 14\ddot{C} + 8\dddot{C} - 2\dot{R} + 6\ddot{R} - 8\dddot{R}$$

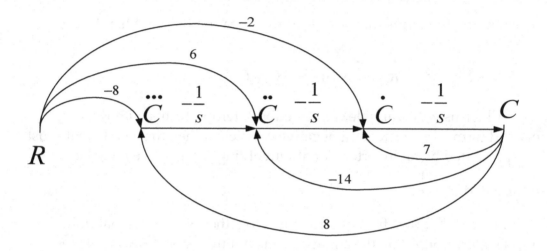

Figure 8.7 Synthesis of a Signal Flow Graph Using the Derivatives of C(s)

$$R(-8)\left(-\frac{1}{s}\right)\left(-\frac{1}{s}\right)\left(-\frac{1}{s}\right) = 8R\left(\frac{1}{s^3}\right)$$

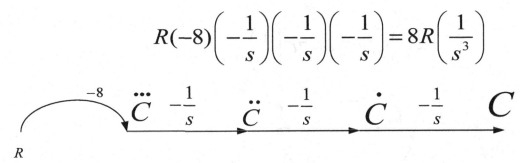

(a) First forward path

$$6R\left(\frac{1}{s^2}\right)$$

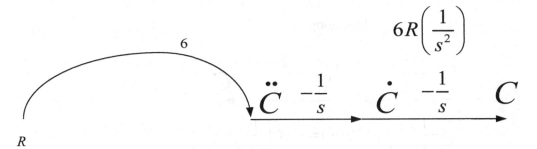

(b) Second forward path

$$2R\left(\frac{1}{s}\right)$$

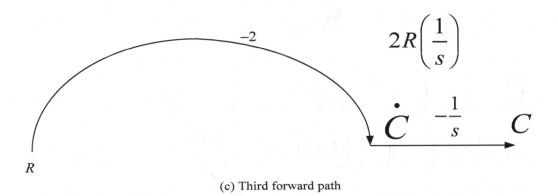

(c) Third forward path

Figure 8.8 Method of verifying the Forward Gains

$$\left(-\frac{1}{s}\right)(C)(7) = -7C\left(\frac{1}{s}\right)$$

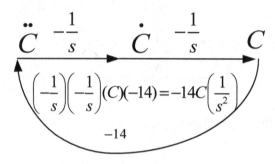

(a) First loop

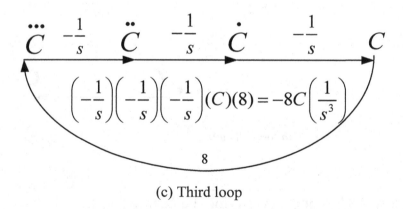

(b) Second loop

(c) Third loop

Figure 8.9 Method of verifying the Loop Gains

$$\frac{C}{R}(s) = \frac{2s^2 + 6s + 8}{s^3 + 7s^2 + 14s + 8}$$

$$C = 7\left(-\frac{1}{s}\right)C - 14\left(\frac{1}{s^2}\right)C + 8\left(-\frac{1}{s^3}\right)C - 2\left(-\frac{1}{s}\right)R + 6\left(\frac{1}{s^2}\right)R - 8\left(-\frac{1}{s^3}\right)R$$

$$C = 7\dot{C} - 14\ddot{C} + 8\dddot{C} - 2\dot{R} + 6\ddot{R} - 8\dddot{R}$$

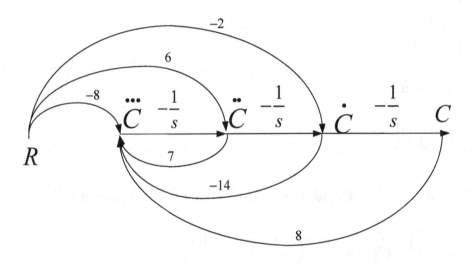

Figure 8.10 Type II Synthesis of the same Transfer Function Shown on Figure 8.7

The above synthesis uses diverging feedback signals from the output. Another type of synthesis uses converging feedback signals from integrators to a summer. Figure 8.11 shows the realization of both types. Components on each diagram were superimposed on the signal flow graph to show how easy it is to design.

## 8.5.3    Proof of the procedure using the derivative of $C(s)$

The procedure above maps $\dot{C}$ to $\left(-\dfrac{1}{s}\right)C$, $\ddot{C}$ to $\left(\dfrac{1}{s^2}\right)C$, and $\dddot{C}$ to $\left(-\dfrac{1}{s^3}\right)C$. Consider the unit step response:

$$c(t) = 1.$$

Its Laplace transform is

$$C(s) = \frac{1}{s}.$$

The first, second, third, and nth derivatives of $C(s)$ are shown below:

$$\frac{dC(s)}{ds} = \dot{C}(s) = \frac{-1}{s^2} = \left(-\frac{1}{s}\right)\frac{1}{s} = \left(-\frac{1}{s}\right)C$$

$$\frac{d^2C(s)}{ds^2} = \ddot{C}(s) = 2\left(\frac{1}{s^2}\right)C$$

$$\frac{d^3C(s)}{ds^3} = \dddot{C}(s) = 6\left(-\frac{1}{s^3}\right)C$$

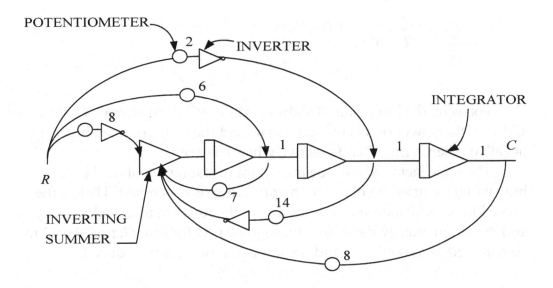

(a) Type I realization of the transfer function

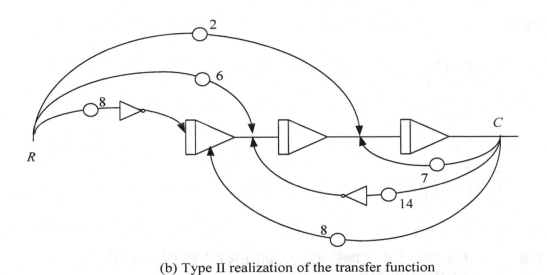

(b) Type II realization of the transfer function

Figure 8.11 Realization of the Transfer Function

$$\frac{d^n C(s)}{ds^n} = \overset{n}{C}(s) = n!\left(\frac{(-1)^n}{s^n}\right)C.$$

Focus on the last result. It shows that the $n^{th}$ derivative of $C$ scales up $C$ by $n!$, the power of $s$ is the same as $n$, and the sign in square bracket is negative when $n$ is odd and positive when $n$ is even.

The scale factor $n!$ may be important in stability analysis but it is immaterial as far as the object of the procedure is concerned. That is, the procedure simply identifies the routing of the forward and feedback signals and the automatically determine the sign of a coefficient. Hence, the $n!$ may be removed as a coefficient and the last equation may be reduced to:

$$\frac{d^n C(s)}{ds^n} = \overset{n}{C}(s) = \left(\frac{(-1)^n}{s^n}\right)C.$$

Thus,

$$\dot{C} = \left(-\frac{1}{s}\right)C$$

$$\ddot{C} = \left(\frac{1}{s^2}\right)C$$

$$\dddot{C} = \left(-\frac{1}{s^3}\right)C.$$

## 8.6      MINIMALLY COMPACT NUMBER OF ELECTRONIC COMPONENTS

The equation for $C$ determines the exact number of components. Specifically, the number of integrators is equal to the highest power of $s$.

Similarly, the number of inverters is the number of negative signs in the equation for C.

A transfer function, however, may be synthesized in two or more ways. In general, the synthesis that gives the least number of components is the one whose feedback and forward signals are not converging to a single point or device. The convergence of two or more signals will require an inverting summer. Hence, a synthesis in which the forward and the feedback signals are diverging provides the minimum number of components.

Synthesizing a transfer function with minimum number of components improves the reliability of a design and decreases its power loss.

# Chapter 9

# Stability and Scaling of a Realization Diagram

As mentioned in the previous chapters, a transfer function must be stable. The Routh-Hurwitz stability criterion is presented here. Additionally, the phase margin of a Bode plot can indicate the stability of a system.

Time scaling speeds up or slows down a system's response. Amplitude scaling prevents the saturation of an amplifier. It also improves the signal to noise ratio of the system.

## 9.1     GENERAL CONSIDERATIONS

Figure 3.2 of chapter 3 shows the transfer function of a system with feedback. That is, if G is the forward block and H is the feedback block, then the transfer function is

$$\frac{C}{R}(s) = \frac{G}{1+GH}.$$

When the denominator, $1+GH$, of the transfer function is zero then $\frac{C}{R}(s)$ becomes infinite. The denominator $1+GH$ becomes zero when $GH =$ -1.

## 9.1.1 Denominator in product of factors form

If the denominator is in the form of product of factors, a factor in the right half of the complex plane will make the transfer function unstable. For example, if

$$\frac{C}{R}(s) = \frac{K}{(s - a + jb)(s - a + jb)}$$

then the poles are:

$$s = a - jb \text{ and } s = a + jb.$$

The real part, $a$, of both pole are on the positive real part axis. Hence, the poles are in the right half plane and the transfer function is unstable.
Its time response from a unit step is:

$$c(t) = Ke^{at} \cos(bt + \phi).$$

Note that the exponential factor $e^{at}$ will grow infinitely.

## 9.1.2 Denominator in sum of terms form

An example of a transfer function whose denominator is in the sum of terms form is:

$$\frac{C}{R}(s) = \frac{K}{s^3 + 6s^2 + 8s + K}.$$

Note that $K$ in the numerator is equivalent to $G$ in the general transfer function

$$\frac{C}{R}(s) = \frac{G}{1+GH}.$$

The following rules determine the stability of a transfer function when inspecting its denominator:

1. If a term in the denominator has a negative coefficient then the transfer function is unstable.

2. If the denominator has a missing term then it can be unstable. It is stable if and only if the remaining terms are all even power of $s$ or all odd power of $s$. Otherwise, it is unstable.

An example of the first case is the denominator $s^3 + 6s^2 - 8s + K$. An example of the second case is $s^3 + 6s^2 + K$ where the term $8s$ is missing. The last example will be stable if it has all even power of $s$ such as $6s^2 + K$.

## 9.2 THE ROUTH-HURWITZ STABILITY CRITERION

Routh-Hurwitz stability criterion establishes the stability of a system by counting the number of poles in the right half of the complex plane. Recall that if a pole exists in the right half of the plane then the system is unstable. The criterion not only detects such a pole but also their number.

The first step in applying the criterion requires creating a matrix similar to the one shown on Figure 9.1. Its first column consists of the descending power of s. Next, the initial entry for the top row consists of coefficients from the denominator of the transfer function. If the highest power of s is an even number then all entries in the row must be the coefficients of s with even power. Otherwise, entries in the row must be the coefficients of s with odd power.

For the next row, follow the same procedure.

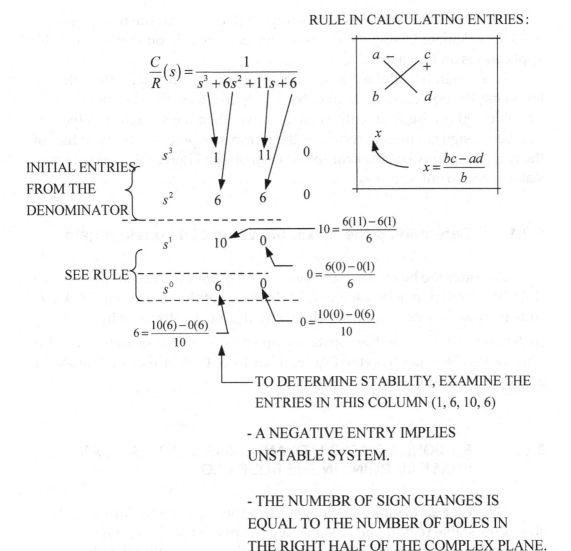

RULE IN CALCULATING ENTRIES:

$$\frac{C}{R}(s) = \frac{1}{s^3 + 6s^2 + 11s + 6}$$

$$x = \frac{bc - ad}{b}$$

INITIAL ENTRIES FROM THE DENOMINATOR

$s^3$    1    11    0

$s^2$    6    6    0

SEE RULE

$s^1$    10    0

$s^0$    6    0

$$10 = \frac{6(11) - 6(1)}{6}$$

$$0 = \frac{6(0) - 0(1)}{6}$$

$$0 = \frac{10(0) - 0(6)}{10}$$

$$6 = \frac{10(6) - 0(6)}{10}$$

TO DETERMINE STABILITY, EXAMINE THE ENTRIES IN THIS COLUMN (1, 6, 10, 6)

- A NEGATIVE ENTRY IMPLIES UNSTABLE SYSTEM.

- THE NUMEBR OF SIGN CHANGES IS EQUAL TO THE NUMBER OF POLES IN THE RIGHT HALF OF THE COMPLEX PLANE.

Figure 9.1 Sample Calculations for Routh-Hurwitz Stability Criterion

After the initial two rows are completed, entries on the remaining rows are calculated using its top two rows. See the rule on Figure 9.1 and its applications on the matrix.

Next, examine all the entries on the second column (after the column involving the powers of s). If an entry is negative then the system is unstable. Otherwise, if all entries are positive, then the system is stable. The number of sign changes determines the number of poles in the right half of the complex plane. For the example shown on the Figure 9.1, the system is stable since no entry is negative.

### 9.2.1     Determining the maximum value of a constant gain

Consider the block diagram shown on Figure 9.2. The Routh-Hurwitz stability criterion may be used to find the range of the constant gain. After constructing the matrix and calculating all the entries, form an inequality such that the entry with the constant gain and in row $s^1$ is greater than zero. This step is also shown on the figure. It has to be greater than zero since all entries in the first column must be positive.

### 9.3     REGIONS OF STABILITY AND INSTABILITY, GAIN AND PHASE MARGINS IN THE BODE PLOT

Figure 9.3 shows a Bode plot of a hypothetical transfer function. Its shows the stable region, the unstable region, phase margin and gain margin. The region to the left of the vertical line intersecting the phase curve at -180

degrees is the stable region. To its right is the unstable region. Designs must not have a phase angle from -180 degrees to 360 degrees.

The figure also shows the rectangle of phase margin and gain margin. It is bounded by the 0 dB vertical line, the -180 degrees vertical line, and frequencies at the top and bottom sides of the rectangle.

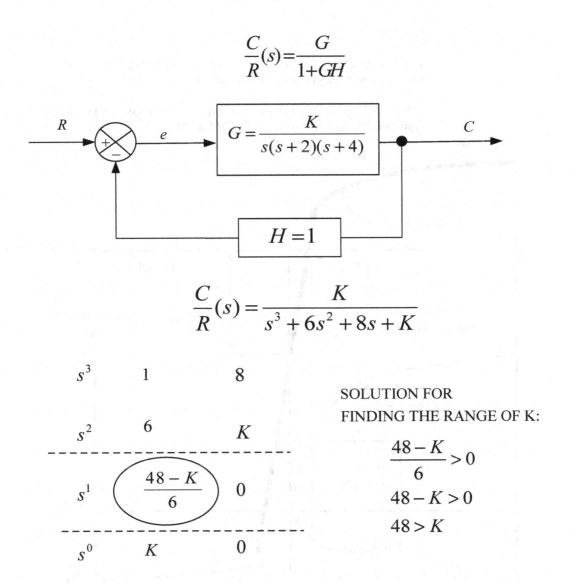

$$\frac{C}{R}(s) = \frac{G}{1+GH}$$

$$G = \frac{K}{s(s+2)(s+4)}$$

$H = 1$

$$\frac{C}{R}(s) = \frac{K}{s^3 + 6s^2 + 8s + K}$$

| | | |
|---|---|---|
| $s^3$ | 1 | 8 |
| $s^2$ | 6 | $K$ |
| $s^1$ | $\frac{48-K}{6}$ | 0 |
| $s^0$ | $K$ | 0 |

SOLUTION FOR
FINDING THE RANGE OF K:

$$\frac{48-K}{6} > 0$$

$$48 - K > 0$$

$$48 > K$$

Figure 9.2 Method of Finding the Range of K

NOTES:

1. THE GAIN MARGIN ALLOWS HOW MUCH MORE OF A GAIN, FROM 0 dB, A SYSTEM CAN TOLERATE.

2. PHASE MARGIN ALLOWS HOW MUCH MORE OF A PHASE, FROM -180 DEGREES, A SYSTEM CAN TOLERATE.

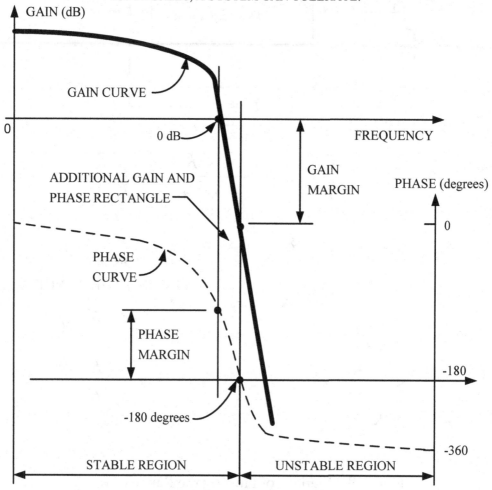

Figure 9.3 Stable and Unstable Regions, and Phase and Gain Margins in a Bode Plot

Essentially, phase margin is the difference between the phase angle when the gain is 0 dB and -180 degrees. Gain margin is the gain between 0 dB and the gain when the phase is at -180 degrees.

Corresponding to the phase margin or the gain margin is the frequency limit for the margin. The limit is controlled by the -180 degrees phase angle.

Large phase or gain margin is desirable.

## 9.3.1 Simplified heuristic proof

Consider the general transfer function

$$\frac{C}{R}(s) = \frac{G}{1+GH}.$$

The gain of the transfer function becomes infinite when its denominator is zero. This occurs when

$$GH = -1.$$

That is,

$$\frac{C}{R}(s) = \frac{G}{1-1}.$$

Factoring the denominator,

$$\frac{C}{R}(s) = \frac{G}{(-1)(-1+1)}.$$

Next, expressing the factors of the denominator in pharos notation,

$$\frac{C}{R}(s) = \frac{G}{\left(|-1|\angle 180^0\right)\left(|0|\angle 0^0\right)}$$

or,

$$\frac{C}{R}(s) = \frac{G}{0\angle 180^0}.$$

Using the phase operation of moving the phase angle from the denominator to the numerator,

$$\frac{C}{R}(s) = \frac{G\angle -180^0}{0}.$$

The result shows that whatever the phase angle of G is, it is shifted by -180 degrees. That is, the condition GH = -1 inverts the output of G. In addition, it creates an infinite gain.

As an example, if H =1, then GH = -1 and G = -1. With this condition, G becomes an inverter of the input signal and no longer depends (on the loose sense) with s. In essence, the RC time constant of the integrator becomes zero (equivalent to having infinite gain).

The result shows that the phase angle of a Bode plot, to be stable, must be between zero degree and -180 degrees. At exactly -180 degrees, the system is marginally stable. Beyond -180 degrees, the system is unstable.

## 9.4 CHANGING THE TIME SCALE OF A REALIZATION DIAGRAM

Time scaling pertains to speeding up or slowing down the response of a transfer function. Its basis is the Laplace transform. Suppose f(t) is the time response and F(s) is the frequency response. The responses are represented by the Laplace transform pair:

$f(t) \leftrightarrow F(s)$

Changing the duration of time by multiplying it by $a$ gives the Laplace transform pair

$$f(at) \leftrightarrow \frac{1}{|a|} F\left(\frac{s}{a}\right).$$

The form of the last pair was from Table 2 of chapter 1.

As an example, suppose $f(t) = e^{-2t}$. Its Laplace transform is $F(s) = \dfrac{1}{s+2}$. Consider next changing the time $t$ by $3t$. That is, $f(3t) = e^{-2(3t)} = e^{-6t}$. Its transform is

$$\frac{1}{|3|} F\left(\frac{s}{3}\right) = \frac{1}{|3|}\left(\frac{1}{\dfrac{s}{3}+2}\right)$$

$$= \frac{1}{s+6}.$$

The result is correct as shown on Table 1 of chapter 1.

Note that when $1 < a < \infty$ the response speeds up. Conversely, when $0 < a < 1$, the response slows down.

Figure 9.3 shows how time scaling applies to the transfer function $\dfrac{C}{R}(s) = \dfrac{1}{s+1}$. It also shows the signal flow graph and realization of the transfer function before and after time scaling.

Two things happen when a transfer function is time scaled. First, all

$s$ in $F(s)$ is replaced by $\dfrac{s}{a}$. Next, the gain is multiplied by $\dfrac{1}{|a|}$. These effects are also shown on the signal flow graph and integrator on Figure 9.4. To synthesize $\dfrac{s}{a}$ is simply change the original value of the capacitance to $\dfrac{C}{a}$. Doing so will also automatically change the gain.

## 9.5    AMPLITUDE SCALING

Amplitude scaling is the process of changing the levels of signals at various points of the system without altering its transfer function. It is usually performed to reduce noise or prevent the saturation of an amplifier. Suppose $G(s)$ has the three factors:

$$G(s) = \left( \frac{10s}{s+1} \right)\left( \frac{10}{s+2} \right)\left( \frac{10}{s+3} \right).$$

Because of the large input in the first factor, the second factor may saturate. To prevent such from happening, the constant gain of the first factor has to be reduced. The gain of the second factor remains. Gain of the third factor has to increase to compensate the reduced gain of the first factor. An example of the result is:

$$G(s) = \left( \frac{s}{s+1} \right)\left( \frac{10}{s+2} \right)\left( \frac{100}{s+3} \right).$$

Note that the overall dynamics of $G(s)$ did not change.

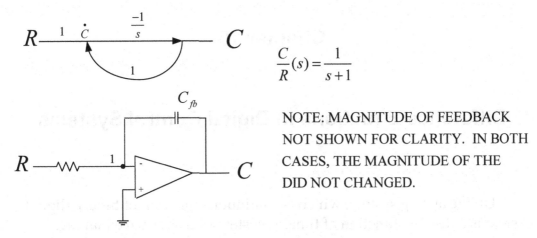

$$\frac{C}{R}(s) = \frac{1}{s+1}$$

NOTE: MAGNITUDE OF FEEDBACK NOT SHOWN FOR CLARITY. IN BOTH CASES, THE MAGNITUDE OF THE DID NOT CHANGED.

(a) Original transfer function, signal flow graph, and circuit realization

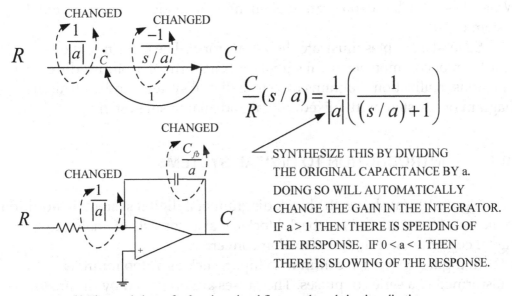

$$\frac{C}{R}(s/a) = \frac{1}{|a|}\left(\frac{1}{(s/a)+1}\right)$$

SYNTHESIZE THIS BY DIVIDING THE ORIGINAL CAPACITANCE BY a. DOING SO WILL AUTOMATICALLY CHANGE THE GAIN IN THE INTEGRATOR. IF $a > 1$ THEN THERE IS SPEEDING OF THE RESPONSE. IF $0 < a < 1$ THEN THERE IS SLOWING OF THE RESPONSE.

(b) Time scaled transfer function, signal flow graph, and circuit realization

Figure 9.4 Application of Time Scaling in a Transfer Function

# Chapter 10

# Difference Equations in Digital Control Systems

Unlike analog system, which is continuous function of time, a digital system is a discrete function of time. A system of difference equations represents a digital system.

The realization procedure of a digital control system is similar to an analog control system. They only differ with the integrating element used. While -1/s is the basic integrating element of an analog system, a digital system uses 1/z.

Several examples illustrate the z-transform. These include the z-transform of common electrical signals, z-transform of arbitrary difference equations, realization diagrams of a pure digital system, and realization diagram of a system with mixed digital and analog subsystems.

## 10.1    INTRODUCTION TO DIGITAL SYSTEMS

Figure 10.1(a) shows the block diagram of a digital system. In addition to the usual forward and feedback blocks, it also consists of analog to digital converter and digital to analog converter.

In digital systems, a continuous input, such as a step input, is transformed to a series of pulses. The pulses are produced by the analog to digital converter whose sampling frequency is at least twice the highest frequency of the signal. While there is limit on the how slow a sample can be made, there is no limit on how fast it could be.

Consider the switch shown on Figure 10.1(b). The input is a continuous signal $y(t)$. Its sampled version is given by the series

NOTES:

1.  THE ANALOG ERROR SIGNAL IS e. ITS SAMPLED VERSION IS e*.

2.  G(z) IS THE z-TRANSFORM OF G.

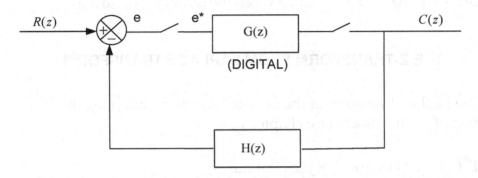

(a) Representation of a digital control system

NOTES:

1.  A/D IS ANALOG TO DIGITAL CONVERTER.

2.  D/A IS DIGITAL TO ANALOG CONVERTER.

3.  AN ACTUATOR CONVERTS ELECTRICAL TO MECHANICAL SIGNAL.
A TRANSDUCER CONVERTS MECHANICAL TO ELECTRICAL SIGNAL.

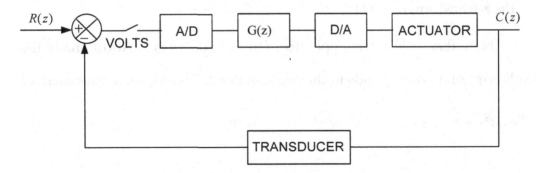

(b) A typical application of digital control system

Figure 10.1 Representations of a Digital Control System and its Typical Application

$$y^*(t) = y_0\delta(t) + y_1\delta(t-T) + y_2\delta(t-2T) + ....$$

The coefficients $y_0, y_1, y_2, ....$ are the amplitudes. Delta functions $\delta(t), \delta(t-T), \delta(t-2T), ...$define when the sampling occurred.

## 10.2     THE Z-TRANSFORM FROM LAPLACE TRANSFORM

The Laplace transform of the sampled signal, by applying the definition of the transform (see chapter 1), is

$$Y^*(s) = y_0(1) + y_1 e^{-Ts} + y_2 e^{-2Ts} + ....$$

Now, let $\dfrac{1}{z} = e^{-Ts}$. Then,

$$Y^*(z) = y_0 + y_1\frac{1}{z} + y_2\frac{1}{z^2} + ....$$

is the z-transform of $y^*(t)$.

Note that $\dfrac{1}{z}$ is a delay operator. The time delay is $T$. Additionally the subscript of $y$ corresponds to the exponent of $z$. That is, the z-transform of $y_{n-2}$ is $y_{n-2}\dfrac{1}{z^{n-2}}$.

## 10.3     Z-TRANSFORM OF COMMON SIGNALS

The z-transform of a step function and a ramp function are shown. These are some of the commonly used electrical signals.

### 10.3.1 Remark on the algebraic representation of signals and its sampled form

Consider the algebraic representation of a ramp such as y = x. The equation means that if x = 0 then y = 0, if x = 1, then y = 1, if x = 2, then y = 2, and so on. These particular values of y are its sampled values. They are used as the coefficients in finding the z-transform of a signal (see $y^*(t)$ and $Y^*(z)$ above). In the following illustrations, the relevant coefficients are represented in the initial equation by parenthesis.

### 10.3.2 z-Transform of a step function

$$Y^*(z) = (1) + (1)\frac{1}{z} + (1)\frac{1}{z^2} + (1)\frac{1}{z^3} + ....$$

$$= \frac{z}{z-1}$$

### 10.3.3 z-Transform of a ramp function

$$Y^*(z) = (0) + (T)\frac{1}{z} + (2T)\frac{1}{z^2} + (3T)\frac{1}{z^3} + ....$$

$$= Tz\left[-\frac{d}{dz}\left(\frac{1}{z-1}\right)\right]$$

$$= \frac{Tz}{(z-1)^2}$$

### 10.4 DIFFERENCE OR RECURSION EQUATION

In a recursion relation, the present value of a function depends on its past value. The equation below is an example of a recursion:

$$y_n + 0.4y_{n-1} + 0.2y_{n-2} = 2x_n + 3x_{n-1}$$

with initial conditions $y_0 = 0$, and $y_1 = 1$; $x$ is a step input.

Solving for $y_n$,

$$y_n = -0.4y_{n-1} - 0.2y_{n-2} + 2x_n + 3x_{n-1}.$$

From the initial conditions and the fact that $x = 1$ (unit step), the other values of $y$'s are:

$$y_2 = -0.4(1) - 0.2(0) + 2(1) + 3(1) = 4.6$$
$$y_3 = -0.4(4.6) - 0.2(1) + 2(1) + 3(1) = 2.96$$
$$y_4 = 2.896$$
$$y_5 = 3.2496$$

Note how the initial conditions and previous calculated value are used to find a present value.

## 10.5 Z-TRANSFORM OF A DIFFERENCE EQUATION AND ITS TRANSFER FUNCTION

Two cases will be shown here. The first case has no initial condition. Case 2 has an initial condition.

### 10.5.1 Case 1 – with no initial condition (or zero initial condition)

Generalize a difference equation as follows:

$$y_n = a_1 y_{n-1} + a_2 y_{n-2} + a_3 y_{n-3} + \dots + b_0 x_n + b_1 x_{n-1} + b_2 x_{n-2} + \dots$$

In digital systems, $y$ represents output. Inputs are the $x$s.

Using the linearity property of the z-transform (see chapter 1),

$$Y(z) = \frac{a_1 Y(z)}{z^1} + \frac{a_2 Y(z)}{z^2} + \frac{a_3 Y(z)}{z^3} + \dots + \frac{b_0 X(z)}{z^0} + \frac{b_1 X(z)}{z^1} + \frac{b_2 X(z)}{z^2} + \dots$$

Collecting similar terms, factoring $Y(z)$ and $X(z)$, and taking their ratio, the transfer function is

$$\frac{Y}{X}(z) = \frac{\dfrac{b_0}{z^0} + \dfrac{b_1}{z^1} + \dfrac{b_2}{z^2} + \dots}{\left(1 - \left(\dfrac{a_1}{z^1} + \dfrac{a_2}{z^2} + \dfrac{a_3}{z^3} + \dots\right)\right)}$$

Simplifying further,

$$\frac{Y}{X}(z) = \frac{\dfrac{b_0}{z^0} + \dfrac{b_1}{z^1} + \dfrac{b_2}{z^2} + \dots}{1 - \dfrac{a_1}{z^1} - \dfrac{a_2}{z^2} - \dfrac{a_3}{z^3} + \dots}.$$

Note the similarity of the result with Mason's rule.

The results above show the important role of indices in quickly determining the z-transform of a difference equation or its transfer function. Specifically, the indices 0, 1, 2, 3, ..., correspond to the coefficients of $b$ and $a$. The coefficients are also the power of z.

## 10.5.2 Case 2 - with initial condition

Suppose the difference equation above has an initial condition $y_0$. The initial condition may be thought of a scaling factor of a delta function. That is,

$$y_0 = a_0 \delta(t).$$

Its z-transform is

$$y_0(z) = a_0(1) = a_0.$$

The result of the case 1 may now be modified to show:

$$Y(z) = a_0 + \frac{a_1 Y(z)}{z^1} + \frac{a_2 Y(z)}{z^2} + \frac{a_3 Y(z)}{z^3} + \ldots + \frac{b_0 X(z)}{z^0} + \frac{b_1 X(z)}{z^1} + \frac{b_2 X(z)}{z^2} +$$

When a difference equation has an initial condition, its transfer function may not be obtained explicitly. It is as if the initial condition is acting as another input. However, when the initial condition is zero, then its transfer function may be found. This is important to realize since if the interest is to find the response of a system, then the initial condition of the output must be zero.

As an example, the z-transform of the previous difference equation (in section 4)

$$y_n + 0.4 y_{n-1} + 0.2 y_{n-2} = 2x_n + 3x_{n-1}$$

with initial conditions $y_0 = 0$, and $y_1 = 1$; $x$ is a step input is

$$Y(z) - 0 = -\frac{0.4Y(z)}{z} - \frac{0.2Y(z)}{z^2} + 2X(z) + \frac{3X(z)}{z^2}$$

or,

$$= -\frac{0.4Y(z)}{z} - \frac{0.2Y(z)}{z^2} + 2X(z) + \frac{3X(z)}{z^2}.$$

## 10.6     TWO SIMULTANEOUS DIFFERENCE EQUATIONS

The following example shows how to solve two simultaneous difference equations. Solve the difference equations

$$y_n = 0.95y_{n-1} + 0.10w_{n-1}$$
$$w_n = 0.95w_{n-1} + 0.05y_{n-1}$$

with initial conditions:

$$y_0 = 20$$
$$w_0 = 40.$$

Finding the z-transform of each equation,

$$Y(z) - 20 = \frac{0.95Y(z)}{z^1} + \frac{0.10W(z)}{z^1}$$
$$W(z) - 40 = \frac{0.95W(z)}{z^1} + \frac{0.05Y(z)}{z^1}.$$

Finally, the solution of $W(z)$ is

$$W(z) = \frac{40z^2 + 37z}{z^2 - 1.9z + 0.8975}.$$

## 10.7      REALIZATION OF A TRANSFER FUNCTION WITH Z-TRANSFORM

The realization of a transfer function in the z-domain follows the same procedure as in the s-domain. That is, after the transfer function is derived, the numerator and denominator are divided by the highest power of z. Then Mason's rule is applied.

Consider the transfer function

$$\frac{Y}{X}(z) = \frac{2z^2 + 2z}{z^2 - 1.25z + 0.375}.$$

After dividing by the highest power of z, the transfer function becomes:

$$\frac{Y}{X}(z) = \frac{2 + \dfrac{2}{z}}{1 - \dfrac{1.25}{z} + \dfrac{0.375}{z^2}}$$

Re-writing the previous result according to Mason's rule,

$$\frac{Y}{X}(z) = \frac{2 + \dfrac{2}{z}}{1 - \left( \dfrac{1.25}{z} - \dfrac{0.375}{z^2} \right)}.$$

As in the s-domain, the forward paths are terms from the numerator. Feedback paths are from the denominator. Figure 10.2(a) is its signal flow graph. The realization diagram is shown on Figure 10.2(b). Note that the

$$\frac{Y}{X}(z) = \frac{2z^2 + 2z}{z^2 - 1.25z + 0.375}$$

$$\frac{Y}{X}(z) = \frac{2 + \dfrac{2}{z}}{1 - \left(\dfrac{1.25}{z} - \dfrac{0.375}{z^2}\right)}$$

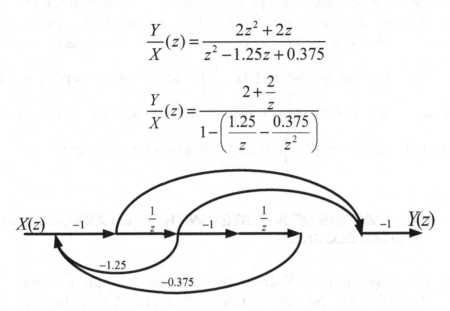

(a) Transfer function and its signal flow graph

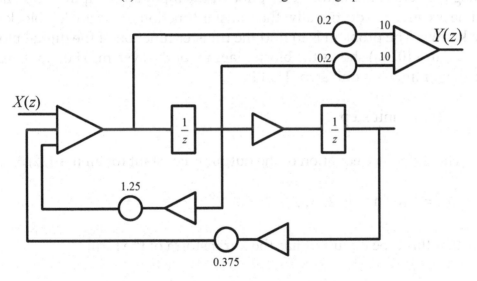

(b) Realization diagram

Figure 10.2 Signal Flow Graph and Realization Diagram of a Digital Transfer Function

sign of a coefficient in a feedback path may be changed so that the product of the signs in the *complete* loop formed by the feedback path is the same as that of the sign in Mason's rule. For example, the sign of 1.25 is positive in Mason's rule. The loop consists of $(-1)$, $\left(\dfrac{1}{z}\right)$, and the feedback path of 1.25.

To make the product of the signs positive, the feedback path must have a negative sign. That is $(-1)\left(\dfrac{1}{z}\right)(-1.25)$ will give a positive sign.

## 10.8 REALIZATION OF A SYSTEM WITH MIXED ANALOG AND DIGITAL BLOCKS

The previous example illustrates the realization of purely digital system. What follows is the combination of digital block (in z domain) and analog block (in s domain). Description of the input and output as function of time are given. Additionally, the transfer function of the analog block is also known. The problem is to find the transfer function of the digital block.

Figure 10.3(a) shows the block diagram of the system. The input, as a function of time is a unit step. That is,

$r(t) = $ unit step.

The difference equation of the output is constant for all n = 1, 2, 3, ...

$c_n = 1$ for n = 1, 2, 3, ....

Note that this type of output has the z-transform of the form

$$\delta_{n-n_0} = \frac{1}{z^{n_0}}.$$

The two switches sample every $T = 2$ seconds.

Taking the z-transform of the input and output,

$$R(z) = \frac{z}{z-1}.$$

$$C(z) = \frac{1}{z} + \frac{1}{z} + \frac{1}{z} + ... + = \frac{1}{z-1}.$$

The corresponding ratio of the output and input is,

$$\frac{C}{R}(z) = \frac{1}{z}.$$

But,

$$\frac{C}{R}(z) = \frac{G(z)}{1+G(z)D(z)}$$

$$= \frac{1}{z}.$$

Using the transformation rule from s to z (see chapter 1),

$$G(s) = \frac{5}{s^2} \leftrightarrow G(z) = \frac{10z}{(z-1)^2}.$$

After some algebra,

GIVEN:

FIND:

$r(t) =$ unit step

$c_n = 1$ for n = 1, 2, 3, ....

$T = 2$ seconds

$G(s) = \dfrac{5}{s^2}$

$D(z)$

$m_n$

SOLUTION:

$D(z) = \dfrac{z-1}{10z} = \dfrac{1}{10} - \dfrac{1}{10}\left(\dfrac{1}{z}\right)$

$m_n = \dfrac{1}{10}e_n^* - \dfrac{1}{10}e_{n-1}^*$

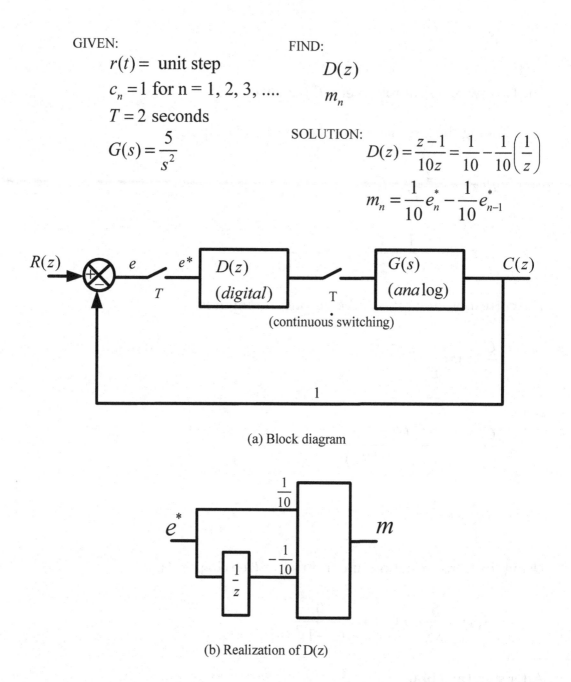

(a) Block diagram

(b) Realization of D(z)

Figure 10.3 System with Mixed Analog and Digital Blocks

$$D(z) = \frac{\left(\dfrac{\dfrac{C}{R}}{1 - \dfrac{C}{R}}\right)}{G}$$

$$= \frac{z-1}{10z} = \frac{1}{10} - \frac{1}{10}\left(\frac{1}{z}\right)$$

The product of the z-transform of $e^*(z)$ and $D(z)$ is $m(z)$. That is,

$$m(z) = e^*(z)\left(\frac{1}{10} - \frac{1}{10}\left(\frac{1}{z}\right)\right)$$

$$= \frac{1}{10}e^*(z) - \frac{1}{10}\left(\frac{1}{z}\right)e^*(z).$$

Taking the inverse z-transform of the last equation gives the difference equation of $m$,

$$m_n = \frac{1}{10}e_n^* - \frac{1}{10}e_{n-1}^*.$$

Figure 10.3(b) shows the realization of $D(z)$.

## 10.8    MICROCONTROLLER IN DIGITAL CONTROL SYSTEMS

Recall that z-transform is a delay operator. It delays a signal by $T$ seconds. A microcontroller can provide such a delay. Normally, software can produce the delay. In other cases, a timer may produce the same.

Other features of a microcontroller include analog inputs, programmable frequency source, analog to digital converter, pulse width modulator, and interrupts. An interrupt stops the execution of a program. The program counter of the microcontroller then jumps to a special program to perform another function such as switching a port.

Note that a typical microcontroller has special function registers that define whether a port is an input or an output. It has also registers that can force the state (high or low) of a port.

Microcontrollers may also be interfaced with digital signal processors that perform mathematical functions (such as discrete Fourier transform). Depending on the result of mathematical calculations, the processor will signal the end of an interrupt and transfers the control back to the microcontroller.

The use of microcontroller in z-transform does not change the realization diagram. It is simply another tool to synthesize delay. For a microcontroller to work in a realization diagram, the output of its clock or equivalent circuit must be in series with an input signal, the sum of input signal and feedback signal, or the sum of a feedback signal and forward signal.

# Appendix A

# Java Program to Calculate Gain and Phase

```java
class QAsymptoticApproximation {

public static void getApproximation() throws IOException {

        Scanner keyboard = new Scanner(System.in);
        FileWriter f = new FileWriter("PowerCalcResults.txt", true);
        PrintWriter pWriter = new PrintWriter(f);
        String M;
        String m0;
        String m1;

        int iNumFreq = 0;

        m0 = "\n\n\t > INPUT the constant in the numerator:
        ";System.out.print(m0);
        double dConstant = keyboard.nextDouble();
        M = m0 + dConstant;
        pWriter.print(M);

        m0 = "\n\n\t > INPUT the number of simple zeros in the numerator:
        ";
        System.out.print(m0);
        int iZero = keyboard.nextInt();
        M = m0 + iZero;
        pWriter.print(M);
```

```
double [] dSimpleZero = new double[iZero];

iNumFreq = iNumFreq + iZero;

m0 = "\n\n\t > INPUT the number of simple poles in the
denominator: ";
System.out.print(m0);
int iPole = keyboard.nextInt();
M = m0 + iPole;
pWriter.print(M);

double [] dSimplePole = new double[iPole];

iNumFreq = iNumFreq + iPole;

m0 = "\n\n\t > INPUT the number of quadratic zeros in the
numerator: ";
System.out.print(m0);
int iQuadZero = keyboard.nextInt();
M = m0 + iQuadZero;
pWriter.print(M);

double [] dCoeffBn = new double[iQuadZero];
double [] dCoeffCn = new double[iQuadZero];

if (iQuadZero > 0) {
iNumFreq = iNumFreq + 2*iQuadZero; // each quad zero takes one
extra cell
}//ei

m0 = "\n\n\t > INPUT the number of quadratic poles in the
denominator: ";
System.out.print(m0);
int iQuadPole = keyboard.nextInt();
M = m0 + iQuadPole;
pWriter.print(M);
```

```java
double [] dCoeffBd = new double[iQuadPole];
double [] dCoeffCd = new double[iQuadPole];

if (iQuadPole > 0) {

iNumFreq = iNumFreq + 2 * iQuadPole;

}//ei

//array for corner frequencies

double [ ] daFreq = new double [iNumFreq];

int p = 0;
int r = 0;
for (int i = 0; i < dSimpleZero.length; i++) {

r = i + 1;
m0 = "\n\n\t > INPUT zero " +r +": ";
System.out.print(m0);
dSimpleZero[i] = keyboard.nextDouble();
M = m0 + dSimpleZero[i] ;
pWriter.print(M);

daFreq [p] = dSimpleZero[i];

p++;

}//ef

for (int i = 0; i < dSimplePole.length; i++) {

r = i + 1;

m0 = "\n\n\t > INPUT pole " +r +": ";
System.out.print(m0);
```

```java
dSimplePole[i] = keyboard.nextDouble();
M = m0 + dSimplePole[i] ;
pWriter.print(M);

daFreq [p] = dSimplePole[i];

p++;

}//ef

for (int i = 0; i < iQuadZero; i++) {

r = i + 1;

m0 = "\n\n\t > INPUT coefficient of S (middle term) in quadratic
zero " +r +": ";
System.out.print(m0);
dCoeffBn[i] = keyboard.nextDouble();
M = m0 + dCoeffBn[i] ;
pWriter.print(M);

m0 = "\n\n\t > INPUT constant term in quadratic zero " +r +": ";
System.out.print(m0);
dCoeffCn[i] = keyboard.nextDouble();
M = m0 + dCoeffCn[i] ;
pWriter.print(M);

daFreq [p] = dCoeffBn[i] / dCoeffCn[i];

p++;

daFreq [p] = Math.sqrt(dCoeffCn[i]);

p++;
```

```java
}//ef

for (int i = 0; i < iQuadPole; i++) {
r = i + 1;

m0 = "\n\n\t > INPUT coefficient of S (middle term) in quadratic
pole " +r +": ";
System.out.print(m0);
dCoeffBd[i] = keyboard.nextDouble();
M = m0 + dCoeffBd[i] ;
pWriter.print(M);

m0 = "\n\n\t > INPUT constant term in quadratic pole " +r +": ";
System.out.print(m0);
dCoeffCd[i] = keyboard.nextDouble();
M = m0 + dCoeffCd[i] ;
pWriter.print(M);

daFreq [p] = dCoeffBd[i] / dCoeffCd[i];

p++;

daFreq [p] = Math.sqrt(dCoeffCd[i]);

p++;

}//ef


// sort the frequencies prior to calculations

m0 = "\n\n\t > INPUT 1 to sort the frequencies, any number to not
sort: ";
System.out.print(m0);
int iSort = keyboard.nextInt();
```

```
M = m0 + iSort;

pWriter.print(M);

if (iSort == 1) {

java.util.Arrays.sort(daFreq);

}// ei

else {

}//ee

m0 = "\n\n\t > TRACE ";
System.out.print(m0);


// header
System.out.print ("\n\n\t Rad/sec \tHertz \t\tGain \t\tdB
\t\tDegrees \n");
pWriter.print ("\n\n\t Rad/sec \tHertz \t\tGain \t\tdB \t\tDegrees
\n");


for (int k = 0; k < iNumFreq; k++) {


// initial values - critical to set before each new frequency
double d1Factor = dConstant;
double d2Factor = 1.0; // critical initialization (can't be zero since it
will give infinity
double dGain = 0.0;
double dRadian = 0.0;
double d1Term = 0.0;
double d2Term = 0.0;
```

```java
double dAngle = 0.0;
double d1Angle = 0.0;

double d2Angle = 0.0;
double dGainLog = 0.0;
double dHz = 0.0;

    dRadian = daFreq[k];

  for (int i = 0; i < dSimpleZero.length; i++) {

    d1Factor = d1Factor * Math.sqrt( (dRadian * dRadian  +
dSimpleZero[i] * dSimpleZero[i] ));
    d1Angle = d1Angle + Math.atan(dRadian / dSimpleZero[i]);

    }

    for (int i = 0; i < iQuadZero; i++) {

        d1Term = dCoeffCn[i] - dRadian * dRadian;
        d1Term = d1Term / dCoeffBn[i];
    d1Factor = d1Factor * Math.sqrt( (dRadian * dRadian + d1Term *
d1Term ));
    d1Factor = d1Factor * dCoeffBn[i];
    d1Angle = d1Angle + Math.atan(dRadian / d1Term);

    }

  for (int i = 0; i < dSimplePole.length; i++) {

    d2Factor = d2Factor * Math.sqrt( (dRadian * dRadian +
dSimplePole[i] * dSimplePole[i] ));
    d2Angle = d2Angle + Math.atan(dRadian / dSimplePole[i]);

    }
```

```
for (int i = 0; i < iQuadPole; i++) {

d2Term = dCoeffCd[i] - dRadian * dRadian;
d2Term = d2Term / dCoeffBd[i];
d2Factor = d2Factor * Math.sqrt( (dRadian * dRadian + d2Term *
d2Term ));
    d2Factor = d2Factor * dCoeffBd[i];
    d2Angle = d2Angle + Math.atan(dRadian / d2Term);

    }


dHz = dRadian / (2.0 * Math.PI);

dGain = d1Factor / d2Factor;
dGainLog = 20.0 * Math.log10(dGain);

dAngle = d1Angle - d2Angle;
dAngle = dAngle * 180.0 / Math.PI ;

dRadian = 1000.0 * dRadian;
dRadian = (int) dRadian;
dRadian = dRadian/1000;

dHz = dHz * 1000.0;
dHz = (int) dHz;
dHz = dHz / 1000;

dGain = 1000.0 * dGain;
dGain = (int) dGain;
dGain = dGain / 1000;

dGainLog = 1000.0 * dGainLog;
dGainLog = (int) dGainLog;
dGainLog = dGainLog / 1000;
```

```java
        dAngle = 1000.0 * dAngle;
        dAngle = (int) dAngle;
        dAngle = dAngle / 1000;


        System.out.print ("\n\t "+ dRadian + "\t\t" +dHz+ "\t\t" +dGain+
        "\t\t" +dGainLog+ "\t\t" +dAngle);
        pWriter.print ("\n\t "+ dRadian + "\t\t" +dHz+ "\t\t" +dGain+
        "\t\t" +dGainLog+ "\t\t" +dAngle);


        }//ef

f.close();

}//em-

    public static void sort(int [] x, int [] y, int [][] z) throws IOException {
        int j = 0;
        int k = 0;
            for (int i = 0; i<x.length-1; i++) {
                    if(x[i] == x[i+1]) {
                    z[j][k] =y[i];
                    k++;
                    }//ei
                    else {
                    z[j][k] =y[i];
                    k = 0;
                    j++;
                    }//ee
                }//ef
        }//em
}//ec-
```
Java is a trademark of Sun Microsystems, Inc.

171

# References

Recommended Handbooks on Electrical Engineering

1. Chen, Wai-Kai, Editor-in chief, The Circuits and Filters Handbook, Massachusetts, CRC Press, Inc., 1995.

2. Reeve, Whittman, D., Subscriber Loop Signaling and Transmission Handbook: Digital, The Institute of Electrical and Electronics Engineers, Inc., New York, NY, 1995.

3. Whitaker, Jerry C., Editor-in-chief, The Electronics Handbook, Massachusetts, CRC Press, Inc., 1995.

References on Mathematics, Complex Numbers, and Partial Differential Equation

4. Abramowitz, Milton, and Stegun, Irene A., editors, Handbook of Mathematical Functions, 9th printing, New York: Dover Publications, 1965.

5. Kreyzig, Erwin, Advanced Engineering Mathematics, New York: John Wiley & Sons, Inc., 1979.

References on the use of Laplace Transform, Aanlog Circuits, Analog Control System, and Digital Control System

6. Carson, Chen, Active Filter Design, New Jersey: Hayden Book Company, 1982.

7. D'Azzo, John J, and Houpis, Constantine, H, Linear Control System Analysis and Design, 2nd edition, New York: McHraw-Hill, 1981.

8. Glasford, Glen M., Analog Electronic Circuits, New Jersey: Prentice-Hall, 1986.

9. Phillips, C. L., and Nagle, H., Jr., Digital Control System Analysis and Design, New Jersey: Prentice-Hall, 1984.

10. Raven, Francis, H, Automatic Control Engineering, New York: McGraw-Hill, 1961, 64-88.

11. Stanley, William D., Operational Amplifiers with Linear Integrated Circuits, New York: Macmillan, 1989.

12. Van Valkenberg, M. E., Network Analysis, 3rd edition, New Jersey: Prentice-Hall, 1974.

References from the Internet

13. Wikipedia contributors, "Bode plot," Wikipedia, The Free Encyclopedia, http://en.wikipedia.org/w/index.php?title=Bode_plot&oldid=2155 57591 (accessed May 29, 2008).

14. Wikipedia contributors, "Laplace transform," Wikipedia, The Free Encyclopedia, http://en.wikipedia.org/w/index.php?title=Laplace_transform&ol did=215593324 (accessed May 29, 2008).

15. Wikipedia contributors, "Transfer function," Wikipedia, The Free Encyclopedia, http://en.wikipedia.org/w/index.php?title=Transfer_function&oldi d=198193742 (accessed May 29, 2008).

# Index